MATH TO BUILD ON

A BOOK FOR THOSE WHO BUILD

JOHNNY E. HAMILTON
MARGARET S. HAMILTON

CONSTRUCTION TRADES PRESS
P.O. BOX 953
CLINTON, NC 28328-0953
919-592-1310

For information contact:

Construction Trades Press
P.O. Box 953
Clinton, NC 28328-0953

(919) 592-1310

Printed in the United States of America

Order form is located in the back of this book.

10 9 8 7 6 5 4 3 2 1

Library of Congress Cataloging in Publication Data

Hamilton, Margaret S., 1952- .
Math to Build On: A Book for Those Who Build/Margaret S. Hamilton, Johnny E. Hamilton.
p. cm.
Includes Index.

ISBN 0-9624197-1-0: $22.95
1. Mathematics. I. Hamilton, Johnny E., 1947-
II. Title.

QA39.2.H315 1993
516.2'02469--dc20

93-10890
CIP

Acknowledgments

Acknowledgments are fun to write because you publicly get to thank people for helping you on a particular project and in your life. They also get you into trouble since some very important people always are left out.

I want to acknowledge the help and encouragement that I constantly get from people on jobs. My boss, **Richard L. Miller**, has gone beyond just being supportive. I thank him for his tremendous backing and patience. The support of **Jeff Bailey**, who preceded Dick as my boss, is also appreciated. The guy who works beside me, **Monte Fillingim**, has made many suggestions that have resulted in changes to parts of this book. Electricians **Victor Arevalo** and **Richard Arizmendi** have provided valuable insights by sharing layout problems from the field. Thank you, **Mack Holland**, for telling me how to square a circle and for sharing your books and experiences. Thanks to **Dwain Bennett** for encouraging me to get this book out for his employees and for his sons. I have received encouragement from many people in my work. Thanks to all of you.

Since publishing the first book, we have received quite a lot of encouragement from people in training programs. It was their comments that caused this book to be written. Thank you **Art**, **Elizabeth**, **Chuck**, **Tom**, **Joe**, **Glenn**, **Joel**, **Mike**, **Jim**, **Richard**, **Chris**, **Dave**, **Eddie**, **Don**, **Gil**, **Mark**, **Ron**, **Larry**, **Scott**, **Carl**, **Charlie**, **Lee**, **Jenny**, and **Vince**.

We have received immense support from some other very important people in our lives — our family. **Kimberly Marchelle Turner** is forever our sunshine. Also, we have four brothers and two sisters who are always patient with us and there for us. Thank you **Billie**, **David**, **Eleanor**, **James**, **Jimmy** and **Ralph**. They, along with **Laurie**, have had pages shoved in their faces to be proofread. Another important person is **Lynne Ingram**. She is called cousin, but fits in the heart more like a sister. Nephews and nieces have contributed their share. **Keely Sutton** reviewed sections and offered suggestions that were used. **Edwin Lamb** gave us a much needed shot in the arm when he reviewed the book during his Thanksgiving holiday. He also took time to review it again this spring. **Gordon** and **Meg Hamilton** have rejuvenated us too many times to count. Thanks also for the support of **Kimbo Sutton** and the **Toboggan Club** and to **Amanda**, **Clarity**, **Felicia**, **Gayle**, **Heather**, **Lena**, **Rachel**, **Ricky**, **Scenia**, **W.C.(Bo)**, and **William.**

Joe Shideler, **Bill Parsons,** and **Mark Rittmann** have been friends of ours for a long time. Friends hold up the mirror for you to see yourself. These guys have held the mirror steady. Thanks also to **Lib** and **Gene Skipper** for the use of "Skippers Quarters".

Leer Larkins held a larger mirror than most could manage. He was quietly beneficial in more ways than most knew. Once when I was without money or food he asked the Medinas of **Zacatecas Restaurant** in **Riverside, California** to feed me until I got back on my feet. They did, so if you are ever in Riverside, please go eat a fine Mexican dinner there and tell them I sent you. You'll have to pay for yours, though. You will no longer find **Leer** at one of the tables. He's gone to see what lies ahead. Good-bye, old friend.

Chris Forhan teaches, writes, and massages words. He is a published poet and our friend. He doesn't just edit our books, he polishes them. We are grateful for the attention he pays to our work.

The artistic influence of **Rebecca Freeman**, **Jim Ann Howard**, **Leer Larkins**, and **Marcel Duchamp** are always felt.

Gibbs Langley always works our books and looks for math mistakes. His input and friendship are treasured.

Last, but not least, is **you**. Your desire to know gives us something to write about. Thank you. See you down the road.

Dedication

This book is dedicated to our parents. They were our "at home" teachers who happened also to be schoolteachers. Combined, they spent over a hundred years in the front of a classroom. We, along with thousands of others, think they did an outstanding job. It is almost impossible to go into town without at least one former student stopping us to make a comment about the effect one of our parents had on his or her life.

Our attitudes about parents and teachers often change as we age and experience life outside their protective umbrella. Shortly after leaving their nest, we are absolutely sure they could have presented us information in a more palatable form; however, as we gain more experience and perhaps children of our own, we wonder how they were able to teach us anything at all.

We now realize that our parents and teachers were just taller and could see further down the road. Their job was to install guideposts and road signs along the way to help make us aware of things that we may not otherwise have noticed.

TO OUR PARENTS AND TEACHERS

DAVID EARL SUTTON & RUBY GORE SUTTON

AND

WILLIAM ERLIE HAMILTON & FRANCES JOHNSON HAMILTON

Preface

Math is the common thread which runs through all design and construction work. The physical work and the techniques used by different trades differ according to those trades' various requirements, but the basic math used doesn't change. Every person, from the amateur wood worker to the most skilled trades craftsman, must be able to determine the exact lengths of the materials for their work. While many skills are not transferable from trade to trade, math skills and the ability to be precise are.

Gertrude Stein said, “A rose is a rose is a rose.” We could say, “An arc is an arc is an arc.” Wherever arcs are used—whether in bends, elbows, curves on tanks, offsets in sidewalks, painted designs on walls, mosaics on floors, curved staircases or logos, they are all calculated through the use of the same formulas. We could also say, “A right triangle is a right triangle is a right triangle.” Right triangles are used to calculate everything from loads on cranes to offsets in molding. Together, the arc and right triangle are the base for all calculations in the trades. From this base, craftpeople can calculate any of their work. Beyond this base is the engineer's realm of calculus and analytical geometry.

It is reported that about 90% of the people in this country feel uncomfortable with their math abilities. Such a large percentage of uncertainty in areas where so much math is used is disastrous. The amount of math that we really use does not fill volumes and does not need to be intimidating. The basic principles can be stated simply, and after that it's a matter of practice. The proof that those principles work is seen when they solve problems in the field.

Table of Contents

Introduction

There is a base of information about math for people who build that remains the same no matter what type of building work they do. This book describes the knowledge that is needed by anyone who designs, builds, fabricates or maintains just about anything. We have worked hard to make the information in this book straightforward and easy to understand. Once you work to understand the concepts, all that is left is the practice.

It took me years to understand the value of practice. Practice puts the information you learn into your long term memory. An example is in the shooting of free throws in basketball. The coach can show you proper shooting techniques, you can read articles on how to improve your shooting skills or watch other people shoot, but only by actually shooting the ball yourself, can you improve your skills.

I find that, with math particularly, working the problems or practices is as important as understanding the idea presented. In this book there are practices after each major idea presented. The answers for all the problems are located in the back of the book. We suggest that you work the first problem of a practice and then check the answer. If your answer is correct, work a few more problems, and check those answers. At this point, if you are having a problem answering correctly, read and work the section again and then work through the practice.

Be sure that you understand each section before you move on to the next. With any foundation, if you want to build upon it and want it to last, it must be stable. If it's not, the structure will be unsteady and there will always be problems. Each section in this book prepares you for the coming sections. Attempting to learn new material when you are unsure of the old steps will cause confusion. If you find yourself confused, retreat, clear up your confusion, and advance again.

Try not to make a section harder than it is. Math is logical, and what you read means just what it says. If you think you already know the material in a section, try working the first problem of the practice. If your answer is correct, you probably do understand, so finish the practice. If you don't answer all of the questions correctly, quickly and without hesitation, study that section.

Throughout this book, you will find suggestions that you return to a particular page and review the information on that page if you are having problem understanding the new information. You will also find reminders to draw thumbnail sketches to help you visualize the calculation you are doing.

There are two images as icons to help you.
First there is a remember icon to help you keep key points in mind.

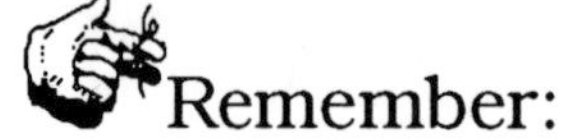

Another symbol you will find is a compass.

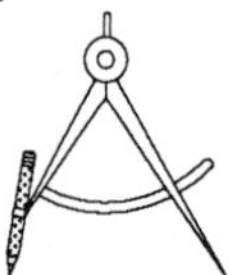

The compass lets you know there is related information in section XVII—Seeing is Believing. This section is a very basic mini-drawing section, which is provided just in case you've forgotten, or never learned, some basic drawing rules. An objective of this book is to encourage you to develop a visual mental image of solutions to problems.

If you are going to buy a scientific calculator to use with this book, there's no need to get fancy. The most inexpensive *scientific* calculator will do. Just make sure that the calculator you buy has the buttons below:

$\boxed{x^2}$ $\boxed{\sqrt{x}}$ $\boxed{\frac{1}{x}}$ $\boxed{\sin}$ $\boxed{\cos}$ $\boxed{\tan}$ $\boxed{\sin^{-1}}$ $\boxed{\cos^{-1}}$ $\boxed{\tan^{-1}}$.

I

Units of Measurement

Units of measurement give meaning to numbers. We speak, write, and calculate numbers with units of measurement every day. We get paid a number of **dollars** and **cents** per **hour.** We measure the **day** in **hours, minutes,** and **seconds**. Length is measured in **miles, yards, feet, inches,** and **fractions of inches.** Carpet measurements are made in **square yards** and concrete and fill dirt are sold in **cubic yards.** Many items in stores are weighed out in **pounds** and **ounces**.

Throughout this book you will constantly be reminded that numbers attached to units of measurement can only be added to or subtracted from numbers attached to the same units of measurement. *(You can't add 1 foot to 1 inch and get 2 inches or 2 feet.)*

Imagine going into a hardware store and saying, "I want a dozen." The clerk would ask, "A dozen what?" You could ask for a dozen till the store closed and the clerk still would not have enough information to help you. A number by itself has little meaning except as a symbol used in counting. To help you, the clerk needs to know the products you want and the unit of measurement for each product.

Suppose you find the items on the list on the previous page, take them to the cash register and hear the clerk say, "Okay, 12 + 4 + 2 + 1 + 1 = 20. You owe $20.00 dollars." You would think the clerk was crazy, yet mixing different units of measurement is a common mistake made in math. Be careful not to make this mistake.

notes

II

Fractions and Decimals

Many different units of measurement are used to express distance (such as the inch, foot, yard, mile, and meter); however, in most trades the inch and the foot are generally used. The accuracy of our work requires that we be able to gauge distances that fall between whole inches. The accuracy needed by a millwright may require that an inch be divided into a thousand or ten thousand equal parts, but in most trades, dividing the inch into 16 equal parts is usually satisfactory. The distance that falls between whole units (in this case, inches) is expressed by what is called a ***fraction***.

Example: $2\frac{1}{2}$" ($2\frac{1}{2}$ inches) falls between 2" and 3"

2 and 3 are the whole numbers
$\frac{1}{2}$ is the fraction

Inch (") is the unit of measurement

A fraction indicates that you are dealing with a part of a whole unit. Just as one quarter ($.25) indicates a part of a whole unit of measurement (the dollar), one quarter of an inch also indicates a part of a whole unit (the inch). In both of the above fractions you are dealing with 1 part of a unit that has been divided into 4 parts ($\frac{1}{4}$).

The bottom number of a fraction is called **the denominator** and **indicates the number of parts the whole unit has been divided into**. The top number, called **the numerator, indicates how many of those parts you are dealing with**. The bar between the top number (numerator) and the bottom number (denominator) of a fraction is a **division bar**.

$$\frac{\text{Numerator} \Rightarrow 1}{\text{Denominator} \Rightarrow 4}$$

For our purposes, **measuring fractions *are*** defined as ***the fractions of a ruler.*** Remember that for most trades, dividing the inch into 16 equal parts is usually satisfactory. If an inch is divided into 16 parts, then the denominator of the fraction is 16. The numerator of the fraction depends upon where (between the whole inches) the measurement ends.

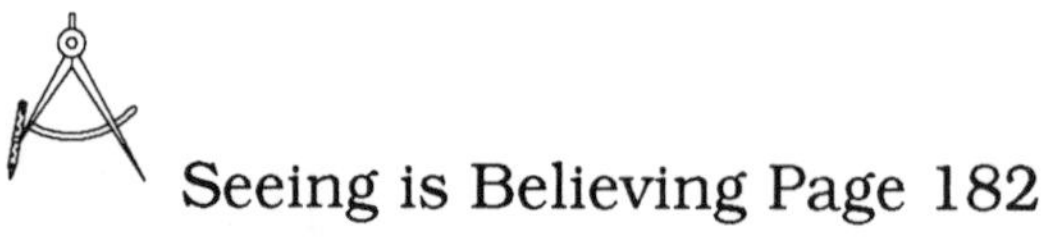
Seeing is Believing Page 182

Denominators

Look at the numbers used as denominators on the fractions below.

1" or $\frac{1}{1}$"	$\frac{1}{2}$"	$\frac{1}{4}$"	$\frac{1}{8}$"	$\frac{1}{16}$"
———————	—————	———	——	-

Notice how each denominator is half as large as the denominator to the right and twice as large as the one to the left. Since the denominator is the number of parts the whole has been divided into, ***the larger the denominator, the smaller a portion of the whole the fraction is.*** You can see this by the lines below the fractions. These lines are the actual length of the fractions. Each line is half as long as the one to the left and twice as long as the one to the right. The larger the denominator, the shorter the line.

Numerators

A whole unit has as many parts as it has been divided into.

For example, if you have ten dimes, then you have *one* dollar. If we express that as a fraction, we would say that ten dimes ($\frac{10}{10}$) equals one dollar ($\frac{1}{1}$).

For every fraction, *if the number of the numerator is the same as the number of the denominator, you have one whole unit.*

Notice that four quarters ($\frac{4}{4}$) equal one dollar and twenty nickels ($\frac{20}{20}$) equal one dollar.

Converting Fractions to Another Term

Converting a fraction to another term is the changing of the numerator **and** *the denominator of a fraction to make a new fraction of the same value.* If the new fraction has smaller numbers, the fraction has been **reduced to a lower term**. If the new fraction has larger numbers, the fraction has been **increased to a higher term**.

Reducing Fractions to Their Lowest Term

Different fractions can be equal to each other. One quarter equals $\frac{1}{4}$ of a dollar, but so does 5 nickels or 25 pennies.

$$\frac{1}{4}=\frac{5}{20}=\frac{25}{100}$$

Look at this chart of the fractions of an inch.

Notice that it is set up just like a ruler.

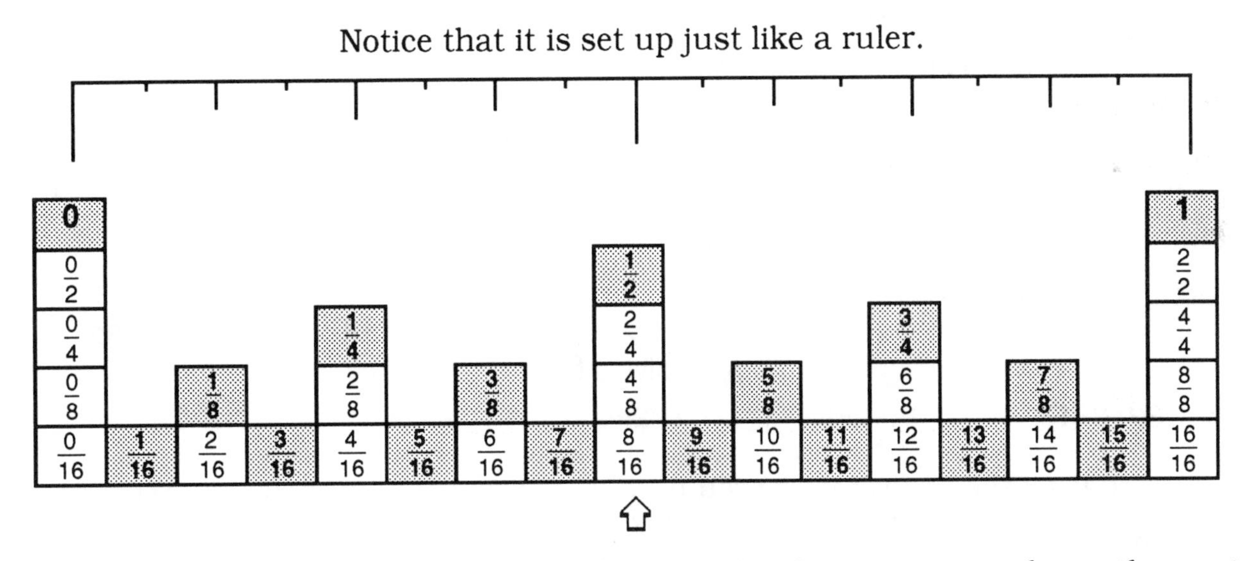

The fractions in each column are equal to each other. For example, in the center column $\frac{1}{2}$, $\frac{2}{4}$, $\frac{4}{8}$, and $\frac{8}{16}$ are all equal to each other.

Just as you have less bulk to carry around when you have a quarter in your pocket instead of 25 pennies, you will have an easier time dealing with fractions when you reduce them to their **lowest term**. The shaded numbers on the chart above are the lowest terms.

To reduce a fraction to a lower term, you must be able to divide the numerator and the denominator by the same number and have the answers for both be whole numbers.

For example, in the fraction $\frac{5}{20}$, both the numerator and the denominator can be divided by 5.

$$\frac{5}{20} = \frac{5 \div 5}{20 \div 5} = \frac{1}{4}$$

$\frac{1}{4}$ is $\frac{5}{20}$ in its lowest term.

Remember:* **The value of a fraction does not change when it is reduced to its lowest term.** One quarter ($\frac{1}{4}$) is equal to five nickels ($\frac{5}{20}$).

Reducing Measuring Fractions to Their Lowest Term

With the measuring fractions (thirty-seconds, sixteenths, eighths, fourths and halves), reduction to a lower term is not difficult. This is the only series of numbers that can be divided by two (2) all the way to one with the result that each answer is still a whole number. Let's take a look.

Start with the number one (1) and multiply it by two (2), then multiply that answer by two and keep going. The answers are the same numbers that we use as the denominators of measuring fractions.

1	2	4	8	16
1 x 2 = **2**	2 x 2 = **4**	4 x 2 = **8**	8 x 2 = **16**	16 x 2 = **32**
		Now divide by 2		
32÷2=**16**	16÷2=**8**	8÷2=**4**	4÷2=**2**	2÷2=**1**

Notice that all the answers are even numbers except the number one. What this means is that with the fractions of a ruler, you will always be able to divide the denominator by 2. ***It's only when the numerator is an odd number that a measuring fraction is in its lowest term.***

When the numerator and the denominator of a fraction are both even numbers, the fraction can always be reduced to a lower term by dividing both by 2.

* You will see this icon and the word, remember, throughout this book to remind you of something you have already studied.

• *If the numerator and the denominator can be divided by more than one number, use the largest possible number as the divisor (divider).*

Example: $\frac{4}{16}$ can be divided by either 2 or 4. In both cases, the answers for both the numerator and the denominator are whole numbers, but you save yourself a step by using the larger number.

first step ⬇ second step ⬇

$$\boxed{\frac{4}{16}} = \frac{4 \div 2}{16 \div 2} = \frac{2}{8} = \frac{2 \div 2}{8 \div 2} = \frac{1}{4}$$

or

One step ⬇

$$\boxed{\frac{4}{16}} = \frac{4 \div 4}{16 \div 4} = \frac{1}{4}$$

Remember: **With measuring fractions, you have only 2, 4, 8, and 16 to choose from as the divisor when reducing to lowest terms.**

Practice 1: Reduce each fraction to its lowest term. Answers-Page 217.

(1) $\frac{2}{16}$

(2) $\frac{4}{8}$

(3) $\frac{6}{16}$

(4) $\frac{20}{32}$

(5) $\frac{12}{16}$

(6) $\frac{2}{4}$

(7) $\frac{14}{16}$

(8) $\frac{2}{8}$

Increasing Fractions to Their Higher Term

Fractions can only be added to or subtracted from other fractions which use the same number as denominator. If the fractions don't use the same number as a denominator, they must be converted to the same number before they can be added or subtracted.

• *To increase a measuring fraction to a higher term, you can multiply the numerator* ***and*** *the denominator by two.*

$\frac{1}{2}(\times\frac{2}{2})\rightarrow$	$\frac{2}{4}(\times\frac{2}{2})\rightarrow$	$\frac{4}{8}(\times\frac{2}{2})\rightarrow$	$\frac{8}{16}$
$\frac{1}{4}(\times\frac{2}{2})\rightarrow$	$\rightarrow$	$\frac{2}{8}(\times\frac{2}{2})\rightarrow$	$\frac{4}{16}$
$\frac{1}{8}(\times\frac{2}{2})\rightarrow$	$\rightarrow$	$\rightarrow$	$\frac{2}{16}$
$\frac{1}{16}(\times\frac{2}{2})\rightarrow$	$\rightarrow$	$\rightarrow$	$\frac{1}{16}$

Here is another method for converting fractions to a higher term:

First: Divide the denominator of the fraction you need by the denominator of the fraction you have.

Second: Multiply *both* the numerator and the denominator of the fraction you have by that answer. The result is a new higher term fraction.

Examples: Convert to a higher term fraction:

$\frac{1}{2}$ to $\frac{\quad}{16}$	$16 \div 2 = \mathbf{8}$	$\frac{\mathbf{8} \times 1}{\mathbf{8} \times 2} = \frac{8}{16}$
$\frac{3}{4}$ to $\frac{\quad}{16}$	$16 \div 4 = \mathbf{4}$	$\frac{\mathbf{4} \times 3}{\mathbf{4} \times 4} = \frac{12}{16}$
$\frac{3}{8}$ to $\frac{\quad}{16}$	$16 \div 8 = \mathbf{2}$	$\frac{\mathbf{2} \times 3}{\mathbf{2} \times 8} = \frac{6}{16}$

Here we are back to those numbers again: 2, 4, 8, or 16.

Practice 2: Convert each fraction to the indicated terms. Answers-Page 217.

(1) $\frac{3}{4} \rightarrow \frac{\quad}{8}$		(5) $\frac{2}{8} \rightarrow \frac{\quad}{32}$
(2) $\frac{15}{16} \rightarrow \frac{\quad}{64}$		(6) $\frac{1}{4} \rightarrow \frac{\quad}{16}$
(3) $\frac{1}{16} \rightarrow \frac{\quad}{32}$		(7) $\frac{7}{8} \rightarrow \frac{\quad}{16}$
(4) $\frac{3}{8} \rightarrow \frac{\quad}{16}$		(8) $\frac{3}{16} \rightarrow \frac{\quad}{32}$

Converting Fractions to a Common Denominator

Only fractions with **common denominators** *can be added to or subtracted from each other.* If you are working with a group of fractions with different denominators, convert the smaller numbered denominators to the largest numbered denominator.

Remember: The larger the denominator, the smaller the portion of the whole unit.

Since you can always divide larger things into smaller parts, the common denominator is usually the denominator with the largest number.

Look at the fractions $\frac{1}{2}$, $\frac{3}{4}$, and $\frac{3}{8}$. The denominators in this group of fractions are 2, 4, and 8. You can divide a half and a fourth into eighths, but you can't make an eighth into a half or a fourth. The common denominator for this group is 8.

Practice 3: Determine the common denominator for each group of fractions. Answers-Page 217.

(1) $\frac{1}{2}$ $\frac{1}{4}$

(2) $\frac{3}{8}$ $\frac{1}{2}$

(3) $\frac{9}{16}$ $\frac{3}{4}$ $\frac{1}{2}$

(4) $\frac{15}{16}$ $\frac{1}{32}$ $\frac{3}{4}$

(5) $\frac{3}{16}$ $\frac{3}{64}$ $\frac{1}{2}$

(6) $\frac{1}{2}$ $\frac{1}{4}$ $\frac{1}{8}$ $\frac{1}{16}$

(7) $\frac{1}{64}$ $\frac{1}{2}$ $\frac{3}{4}$ $\frac{1}{4}$

(8) $\frac{15}{16}$ $\frac{3}{16}$ $\frac{9}{16}$ $\frac{1}{64}$

Addition and Subtraction of Fractions

Once the group of fractions you are adding or subtracting has a common denominator, *you add or subtract only the numerators.*

Remember: The numerator is the number of parts that you are dealing with.

For example: $\frac{1}{4}+\frac{1}{4}=\frac{2}{4}$ $\quad \frac{3}{16}-\frac{1}{16}=\frac{2}{16}$ $\quad \frac{3}{8}+\frac{3}{8}=\frac{6}{8}$

Since **only** the numerators are added or subtracted, you can also do it this way: $\frac{1+1}{4}=\frac{2}{4}$ $\quad \frac{3-1}{16}=\frac{2}{16}$ $\quad \frac{3+3}{8}=\frac{6}{8}$

If you add in columns, you must still remember to add or subtract just the numerators.

$$\begin{array}{r} \frac{1}{4} \\ +\frac{1}{4} \\ \hline \frac{2}{4} \end{array} \qquad \begin{array}{r} \frac{3}{16} \\ -\frac{1}{16} \\ \hline \frac{2}{16} \end{array} \qquad \begin{array}{r} \frac{3}{8} \\ +\frac{3}{8} \\ \hline \frac{6}{8} \end{array}$$

Practice 4: Add and/or subtract each group of fractions. Reduce to lowest terms. Answers-Page 217.

(1) $\frac{1}{2} + \frac{1}{2}$

(2) $\frac{1}{2} + \frac{1}{2} + \frac{1}{4}$

(3) $\frac{3}{16} - \frac{1}{16}$

(4) $\frac{3}{8} + \frac{3}{16} - \frac{1}{2}$

(5) $\frac{9}{32} + \frac{3}{8} + \frac{1}{64}$

(6) $\frac{12}{16} - \frac{3}{8}$

(7) $\frac{1}{2} + \frac{15}{16} + \frac{7}{8}$

(8) $\frac{1}{2} + \frac{1}{16} + \frac{1}{64}$

Multiplication and Division of Fractions

Multiplication of Fractions

When you multiply fractions, multiply numerator by numerator and denominator by denominator. There is no conversion necessary.

$$\frac{\text{numerator}}{\text{denominator}} \frac{\text{x}}{\text{x}} \frac{\text{numerator}}{\text{denominator}} \text{ or } \frac{\text{numerator x numerator}}{\text{denominator x denominator}}$$

Example: Multiply $\boxed{\frac{3}{4} \times \frac{5}{8}}$ $\frac{3 \times 5}{4 \times 8} = \mathbf{\frac{15}{32}}$

$\boxed{\frac{1}{2} \times \frac{3}{4}}$ $\frac{1 \times 3}{2 \times 4} = \mathbf{\frac{3}{8}}$

Practice 5: Multiply these fractions. Answers-Page 217.

(1) $\frac{1}{4} \times \frac{1}{2}$

(2) $\frac{3}{4} \times \frac{3}{4}$

(3) $\frac{9}{16} \times \frac{1}{3}$

(4) $\frac{2}{5} \times \frac{1}{3}$

(5) $\frac{5}{8} \times \frac{11}{16}$

(6) $\frac{9}{16} \times \frac{1}{2}$

(7) $\frac{1}{12} \times \frac{3}{8}$

(8) $\frac{3}{32} \times \frac{2}{32}$

Division of Fractions

A fraction is divided by a fraction using a process called inverting and multiplying. You **invert** *a fraction by turning it over.*

$\frac{1}{2}$ becomes $\frac{2}{1}$ and $\frac{3}{4}$ becomes $\frac{4}{3}$.

• To *divide a fraction by a fraction, invert the dividing fraction, then multiply it by the fraction you want to divide.*

Example: Divide $\frac{3}{2}$ by $\frac{1}{2}$ $\quad \frac{3}{2} \times \frac{2}{1} = \frac{3 \times 2}{2 \times 1} = \frac{6}{2} = \mathbf{3}$ There are three $\frac{1}{2}$'s in $\frac{3}{2}$.

Divide $\frac{5}{8}$ by $\frac{1}{4}$ $= \frac{5 \times 4}{8 \times 1} = \frac{20}{8} = 2\frac{4}{8} = \mathbf{2\frac{1}{2}}$

Practice 6: Divide these fractions. Answers-Page 217.

(1) $\frac{9}{16}$ by $\frac{1}{2}$ (4) $\frac{15}{16}$ by $\frac{3}{16}$

(2) $\frac{7}{8}$ by $\frac{1}{4}$ (5) $\frac{13}{16}$ by $\frac{1}{8}$

(3) $\frac{12}{16}$ by $\frac{3}{4}$ (6) $\frac{3}{16}$ by $\frac{1}{32}$

Multiplying and Dividing Fractions by One

If you look back at the fraction chart on page 5, you see that $1 = \frac{1}{1} = \frac{2}{2} = \frac{4}{4} = \frac{8}{8} = \frac{16}{16}$. When a fraction is multiplied or divided by any of these numbers, it is actually being multiplied or divided by one. Notice that this is what you did in converting fractions to a different term. You multiplied (or divided) the numerator *and* the denominator by 2.

Multiplying or dividing a number by 1 does not change the value of the number. You change the numbers of the fraction, but the value of the fraction does not change.

For example: $\frac{1}{4} \times \mathbf{\frac{2}{2}} = \frac{2}{8}$

Improper Fractions

Yes, even fractions can be improper. A fraction is a portion of a whole. *If the numerator of a fraction is the same or larger than the denominator, the fraction is equal to or greater than the whole unit, thus improper.* Two examples of **improper fractions** are $\frac{3}{2}$ and $\frac{7}{4}$. In both cases, the numerator is larger than the denominator. Think of the money in your pocket. If you have three half dollars ($\frac{3}{2}$), then you have a dollar and a half ($1\frac{1}{2}$ or $1.50). If you have seven quarters ($\frac{7}{4}$), then you have a dollar and three quarters ($1\frac{3}{4}$ or $1.75).

Examples: $\frac{5}{4}, \frac{4}{4}, \frac{18}{16}, \frac{3}{2}$

Just because fractions are improper doesn't mean they aren't useful. When whole numbers are converted to fractions, they are improper fractions.

Multiplying Whole Numbers and Fractions

To multiply a whole number by a fraction, you must make the whole number into a fraction — an improper fraction. For example, 6 becomes $\frac{6}{1}$. You can do this because **a number divided by one is the number itself**. Six divided by one is six. Once the whole number is converted to a fraction, you can multiply the fractions and reduce the answer to its lowest term.

$$\boxed{6 \times \frac{3}{4} =} \quad \frac{6}{1} \times \frac{3}{4} = \frac{6 \times 3}{1 \times 4} = \frac{18}{4} = 4\frac{2}{4} = \mathbf{4\frac{1}{2}}$$

The number $4\frac{1}{2}$ is a **mixed number** because it has both a whole number (4) and a fraction ($\frac{1}{2}$).

When multiplying a whole number by a mixed number:

First: Multiply the whole numbers.

Second: Multiply the fraction by the whole number.

Third: Add the two results to get the answer.

Example: $4 \times 2\frac{1}{2}$

$4 \times 2 = \mathbf{8}$	Multiply the whole numbers
$\frac{4}{1} \times \frac{1}{2} = \frac{4}{2} = \mathbf{2}$	Multiply the fraction by the whole number
$8 + 2 = \mathbf{10}$	Add the results together

Practice 7: Multiply each whole number by the fraction or mixed number. Reduce the answers to lowest terms. Answers-Page 217.

(1) $2 \times \frac{1}{2}$

(2) $12 \times \frac{5}{8}$

(3) $7 \times \frac{1}{16}$

(4) $3 \times 3\frac{2}{3}$

(5) $10 \times 2\frac{3}{4}$

(6) $16 \times 8\frac{5}{16}$

Remember: **When you multiply fractions, you multiply them straight across.**

$\left(\frac{\text{numerator x numerator}}{\text{denominator x denominator}}\right)$

Decimal Fractions

When you see a number with a decimal point (.) in front, this should indicate the same thing to you as seeing any other fraction (proper fraction, that is). It means that you are dealing with a portion of a whole unit. The name decimal fraction is usually shortened to just **decimal**. Decimal fractions come in two forms — with a decimal point or with a denominator. Here is the same fraction in both forms.

.1234

Decimal Point
1st Decimal Place
2nd Decimal Place
3rd Decimal Place
4th Decimal Place

$$\frac{1,234}{10,000} = \frac{\text{numerator}}{\text{denominator}}$$

Notice that there are 4 decimal places in the decimal form and 4 zeros in the denominator of the regular fraction form. This is not by chance. *The decimal point takes the place of the division bar and the denominator. The number of zeroes in the denominator indicates the number of decimal places in the decimal form.* Since the decimal point format is easier to deal with, it is the form most often used.

Example: Convert .375 to a regular fraction format.

Write $\frac{.375}{1}$ • Draw a line under the decimal number and write a 1 under the line.

$\frac{375}{1000}$ • Add one zero for each decimal place. Erase the decimal point.

NOTE: Watch out for the zeros when converting from a decimal point format to a fraction format. Those zeros seem to throw some people off.

Example: When converting a number like .0001 to a regular fraction format, you still follow the same rules.

Write $\frac{.0001}{1}$

- Draw a line under the numbers and write a 1 under the line.

$\frac{0001}{10000}$

- Follow the 1 with a zero for each decimal place. Erase the decimal point.
- Erase the zeros (in the numerator) that are in front of the other numbers.

$\frac{1}{10000}$ reads much easier than $\frac{0001}{10000}$

Practice 8: Convert each decimal fraction from a decimal point format to a regular fraction format. Answers-Page 217.

(1) 0.875

(2) 0.4

(3) 0.005

(4) 0.03

(5) 0.1507

(6) 0.014

NOTE: In many examples throughout this book, you will see nine decimal places. That is the number of decimal places shown in the display of many scientific calculators. One ten thousandths of an inch is generally accurate enough, so when using a calculator while working in this book, round your final answers off to four decimal places.

An Interesting Note: When our government was being formed, Thomas Jefferson and others tried to get the decimal system adopted as our standard of measurement. They met a great deal of opposition similar to the opposition to the metric system today. Jefferson's group was successful only in getting the decimal system used in our money system.

The fractional system we use in measurement is the same system used in colonial times as a part of the English system of measurements. Even the British changed to the simpler metric system over twenty five years ago. We are the only major country in the world that still uses the more difficult system.

There is a section on metric measurements at the end of this book.

Converting Fractions to Decimals

Fractions are used when measuring distances in the field. In order for these measurements to be used in calculations, they must be converted to decimals. After the calculations produce an answer, fractions are again needed to apply those measurements in the field, so the decimals are converted back to fractions. The conversions are necessary because of the differences between the measuring tools and the calculating tools.

We convert fractions and decimals every day since our money system is based on the decimal system.

$\frac{1}{2}$ of a dollar is equal to \$0.50 (a half dollar).
$\frac{1}{4}$ of a dollar is equal to \$0.25 (a quarter).
$\frac{1}{8}$ of a dollar is equal to \$0.125 ($12\frac{1}{2}$cents).
$\frac{1}{10}$ of a dollar is equal to \$0.10 (a dime).

Just as fractions of a dollar are converted into decimals, fractions of an inch are converted into decimals:

$\frac{1}{2}$ of an inch is equal to 0.50".
$\frac{1}{4}$ of an inch is equal to 0.25".
$\frac{1}{8}$ of an inch is equal to 0.125".
$\frac{1}{10}$ of an inch is equal to 0.10".

Converting Fractions of Inches to Decimals

Remember: The bar between the top number (numerator) and the bottom number (denominator) of a fraction is a **division bar**.

Division bar ➔ $\frac{\textbf{numerator}}{\textbf{denominator}}$

All you have to do **to convert a fraction to a decimal** is **divide the numerator by the denominator**. The answer is the decimal which is equal to that fraction.

$$\text{Denominator}\overline{\smash{)}\text{Numerator}}^{\text{Decimal}}$$

$$16\overline{)9}\quad .5625$$

Example: $\frac{9}{16}$"

Enter 9 [÷] 16 [=] Display [0.5625]

There is a conversion chart on page 211, but you will save yourself a lot of time if you learn to do conversions on the calculator.

Practice 9: Convert each fraction of an inch to a decimal format. Answers-Page 217.

(1) $\frac{3}{8}$"	(4) $\frac{1}{2}$"	(7) $\frac{7}{8}$"	(10) $\frac{1}{4}$"
(2) $\frac{11}{16}$"	(5) $\frac{3}{4}$"	(8) $\frac{3}{16}$"	(11) $\frac{15}{16}$"
(3) $\frac{7}{16}$"	(6) $\frac{19}{32}$"	(9) $\frac{1}{8}$"	(12) $\frac{1}{16}$"

Converting Mixed Numbers to Decimals

To convert a **mixed number** (*a whole number and a regular fraction*) to a whole number and decimal:

First: Convert the fraction to a decimal.
Second: Add the results to the whole number.

Example: $7\frac{9}{16}$" is a mixed number.

9 ÷ 16 = 0.5625" (convert the fraction to a decimal)

.5625" + 7" = 7.5625" (add the decimal to the whole number)

Practice 10: Convert each mixed number to a whole number with decimals. Answers-Page 217.

(1) $19\frac{1}{2}$"	(3) $3\frac{15}{16}$"	(5) $17\frac{1}{8}$"	(7) $7\frac{1}{32}$"
(2) $43\frac{11}{16}$"	(4) $5\frac{5}{16}$"	(6) $37\frac{7}{8}$"	(8) $24\frac{3}{8}$"

Converting a Fraction of a Foot to a Decimal

Converting Whole Inches to a Decimal of a Foot

Many times measurements taken in the field are in *feet* and inches, rather than just inches. In these cases, the foot is the whole unit of measurement. The inches are the fraction that needs to be converted to a decimal.

Remember: The denominator of a fraction is the number of parts a unit is divided into.

There are 12 inches to a foot, so 12 is the denominator of the fraction. The numerator indicates the number of parts you are dealing with; therefore, the numerator is the number of inches in the measurement.

Example: Convert the whole inches to a fraction of a foot. Then convert the mixed numbers to decimal numbers.

Remember: The bar between the numerator and denominator is a division bar.

Measurement		Fraction		Decimal
9'3"	=	$9\frac{3}{12}$'	=	9.25'
2'6"	=	$2\frac{6}{12}$'	=	2.5'
10' 11"	=	$10\frac{11}{12}$'	=	10.9167'*

Practice 11: First, create a mixed number using the regular fraction form, then convert each measurement to a decimal format. Round off each answer to the fourth decimal place. Answers-Page 217.

(1) 1' 3"
(2) 7' 1"
(3) 62' 10"
(4) 29' 2"
(5) 44' 4"
(6) 137' 5"

* In the case of this decimal, I have rounded the decimal off to 4 places. This is generally accurate enough in the trades. Rounding a number off is done by looking at the number to the right of the number to be rounded off. If the number to the right is 5 or higher, then the round off number is made one number larger. If the number to the right is less than 5, then the round off number stays the same. In both cases, the round off number is the last number in the series. In the example shown above, 10.91666666 rounds off to 4 decimal places to become 10.9167. The 6 is the fourth decimal place and the number to the right of it is a 6, so the fourth decimal place is rounded up to 7 and all of the numbers to the right of the fourth decimal place are dropped.

Converting Inches and a Fraction of an Inch to a Decimal of a Foot

Converting a mixed number with feet, inches, and a *fraction of an inch* to whole feet and a decimal of a foot requires an extra step. You have to convert the fraction of an inch to a decimal of an inch.

Look at the example below to see what a fraction of a foot that includes a fraction of an inch looks like.

$$5'\,4\tfrac{3}{8}" = 5\,\frac{4\frac{3}{8}}{12}' = 5\,\frac{4.375}{12}'$$

Here is a series of steps for the conversion.

First: Subtract the whole feet and inches from the number.

Second: Convert the fraction of an inch to a decimal of an inch.

Third: Add the whole inches to the decimal of an inch.

Fourth: Divide the new number by 12 (the number of inches in a foot).

Fifth: Add in the whole feet.

Example: $73'\,7\,\frac{13}{16}"$ expressed as a fraction of a foot: $73\,\frac{7\frac{13}{16}}{12}'$

$73'\,7\,\frac{13}{16}" - 73'\,7"$	$\frac{13}{16}"$
$13 \div 16 = 0.8125"$	*Divide the fraction of the inch.*
$7" + 0.8125" = 7.8125"$	*Add in the whole inches.*
$7.8125" \div 12" = 0.6510'$	*Divide by 12 (number of inches in a foot).*
$73 + 0.6510 =$ **73.6510'**	*Add the whole feet to the decimal.*

Here is the same example using the calculator.

Enter 13 [÷] 16 [=]	Display	0.8125
[+] 7 [=]	Display	7.8125
[÷] 12 [=]	Display	0.651041666
[+] 73 [=]	Display	73.65104167

73.6510' is the answer after rounding off.

Practice 12: Convert each measurement to a decimal of a foot. First, create a fraction. Round each answer off to 4 decimal places. Answers-Page 218

(1) 5' 6"

(2) 2' 9"

(3) 3' 7 $\frac{3}{4}$"

(4) 13' 8 $\frac{3}{16}$"

(5) 2' 4 $\frac{1}{8}$"

(6) 283' $\frac{5}{8}$"

(7) 42' 8 $\frac{7}{16}$"

(8) 1' 11 $\frac{13}{16}$"

(9) 11' 1 $\frac{1}{4}$"

(10) 87' 6 $\frac{15}{16}$"

notes

Converting Decimals to Fractions

In the field, calculations are made to obtain a needed measurement. Decimals are used in the calculations, but measurements are made in fractions, so you have to convert your final answers from decimals to fractions in order to apply the measurements.

Converting a Decimal of an Inch to a Fraction

To convert a decimal of an inch to a fraction of an inch, simply reverse the method that was used in the last section. **Multiply the decimal of an inch by the denominator that you want to use.**

Denominator x Decimal = Numerator

Since the accuracy generally needed in the trades is based on $\frac{1}{16th}$ of an inch, 16 is the most commonly used denominator. If more accuracy is needed, 32 can be used. The nearest whole number of the answer is the numerator.

Example:	.375	(the decimal of an inch)
	x 16	(the denominator)
	6.000	(the numerator)

You now have the fraction $\frac{6}{16}$", which can be reduced to $\frac{3}{8}$".

When working problems, seldom do you end up with a decimal that will convert to an even sixteenth. You must round off the answer to the nearest whole number. A decimal such as 0.6327 multiplied by 16 will yield an answer of 10.1232.

$$.6327 \times 16 = 10.1232$$

This answer (10.1232) is the numerator, but to be useful to us, it must be rounded off to 10. This produces the fraction $\frac{10}{16}$, which reduces to $\frac{5}{8}$.

Remember: When rounding off to the nearest whole number, if the decimal is .5 or more go up to the next whole number. If the decimal is below .5, use the existing whole number. For example: 3.49 will remain 3, while 3.5 will round off to 4.

Practice 13: Convert each decimal number to the closest sixteenth fraction. Reduce when necessary. Answers-Page 218.

(1) .5486"

(2) .7361"

(3) .3023"

(4) .9999"

(5) .8315"

(6) .1285"

(7) .8397"

(8) .3275"

Converting Inches with Decimals to Inches and Fractions

To convert inches with decimals to inches and fractions:

First: Subtract the whole number from the inches and decimal of an inch.

Second: Convert the decimal of an inch to a fraction of an inch.

Third: Add the whole number to the fraction.

It helps to write the numbers down while doing conversions.

Example: 9.1875"

Write down

First: 9.1875" - **9**" = 0.1875" — *9*

Second: 0.1875 x **16** = **3** — $\frac{3}{16}$

Third: $9'' + \frac{3}{16}'' = \mathbf{9\frac{3}{16}}''$ — $9\frac{3}{16}''$

Practice 14: Convert each whole number with decimal to a whole number with fraction. Use sixteenths as the denominator. Reduce to the lowest term when necessary. Answers-Page 218.

(1) 12.375"

(2) 4.6"

(3) 99.99"

(4) 5.77"

(5) 84.1725"

(6) 12.5737"

(7) 121.839"

(8) 23.892"

Converting a Decimal of a Foot to Inches and a Fraction of an Inch

To convert a number with feet and a decimal of a foot to a number with feet, inches, and a fraction of an inch:

First: Subtract out the number of whole feet.

Second: Multiply the decimal of a foot by 12 (the number of inches in a foot). The whole number in the answer is the inches.

	✎ ***Write down***		
6.75'- 6' = 0.75	6'	= Whole feet	(Subtract out the whole feet)
0.75 x 12" = 9"	+9"	= Whole inches	(Multiply the decimal of a foot by 12)
	6'9"	= Feet and inches	(The whole number is the inches)

When you multiply a decimal of a foot by 12 and the answer is a whole number *only*, you have finished the conversion. If the answer is a whole number and a decimal, the whole number is the number of whole inches and the decimal number needs to be converted to a fraction of an inch.

To change the decimal of an inch to a fraction of an inch:

First: Subtract out the whole number.

Second: Multiply the decimal by 16 (or the denominator of your choice).

Third: Add the feet, the whole inches and the fraction of an inch together.

Example: Convert this number to feet, inches, and fractions.

	✎ Write down	
73.65104' - **73'** = 0.65104'	**73'**	Whole feet
0.65104 x 12 = 7.81248" - **7"** = 0.81248	**7"**	Whole inches
0.81248 x **16** =12.9996 (round off to **13**)	$\mathbf{\frac{13}{16}}$"	Fraction of an inch
Total	**73' 7$\mathbf{\frac{13}{16}}$"**	

Practice 15: Convert each whole feet with decimal to feet, inches, and a fraction of an inch. Reduce the fractions when necessary. Answers-Page 218.

(1) 12.75'
(2) 1.9'
(3) 3.333'
(4) 155.863'
(5) 50.125'
(6) 40.29'
(7) 33.314'
(8) 75.13'
(9) 84.441'
(10) 9.567'
(11) 11.298'
(12) 14.997'

notes

Memory Aid

Doubling a Measuring Fraction

To double a fraction, you normally multiply it by 2, then reduce the answer to the lowest denominator.

Example: $\boxed{2 \times \frac{3}{8}} = \frac{2}{1} \times \frac{3}{8} = \frac{2 \times 3}{1 \times 8} = \frac{6}{8} = \frac{3}{4}$

A quicker way is to divide the denominator by 2.

Example: $\boxed{2 \times \frac{3}{8}} = \frac{3}{8 \div 2} = \frac{3}{4}$

Of course, this works only with fractions that have an even number for the denominator, but our measuring fractions are all based on even numbers. (Examples: $\frac{1}{2}, \frac{1}{4}, \frac{1}{8}$, and $\frac{1}{16}$.)

Halving a Measuring Fraction

To divide a fraction in half, you were told in school and on page 11 to invert and multiply. The shortcut is just to multiply the denominator by 2.

To find half of these fractions, multiply the denominator by 2.

$\boxed{\frac{3}{8} \div 2} = \frac{3}{8 \times 2} = \frac{3}{16}$

$\boxed{\frac{7}{8} \div 2} = \frac{7}{8 \times 2} = \frac{7}{16}$

$\boxed{\frac{3}{4} \div 2} = \frac{3}{4 \times 2} = \frac{3}{8}$

$\boxed{\frac{5}{8} \div 2} = \frac{5}{8 \times 2} = \frac{5}{16}$

Halving Mixed Numbers

A few years ago I learned a new trick from an old Bulldog, Bulldog Holland, on how to divide a mixed number in half. He separated mixed numbers into two categories. The first category was mixed numbers with even whole numbers. The second category was mixed numbers with odd whole numbers.

Mixed Numbers with an Even Whole Number

For example, $32\frac{7}{8}$ is a mixed number with an even whole number (32).

In this case, divide the whole number by 2: $\frac{32}{2} = 16$

Use the short cut for dividing a fraction: $\frac{7}{8 \times 2} = \frac{7}{16}$

Add them back together: $16\frac{7}{16}$

Half of $32\frac{7}{8}$ is $16\frac{7}{16}$

Mixed Numbers with an Odd Whole Number

$11\frac{3}{8}$ is a mixed number with an odd whole number (11).

Divide the whole number by 2: $\frac{11}{2} = 5.5$ (Drop the decimal to make it a whole number of 5).

Add the numerator to the denominator: 3 + 8 = 11 (11 is the new numerator).

Double the denominator: 8 x 2 = 16 (16 is the new denominator).

When you put the two together, the fraction is $\frac{11}{16}$.

Your final answer is $5\frac{11}{16}$.

This drawing shows how it works.

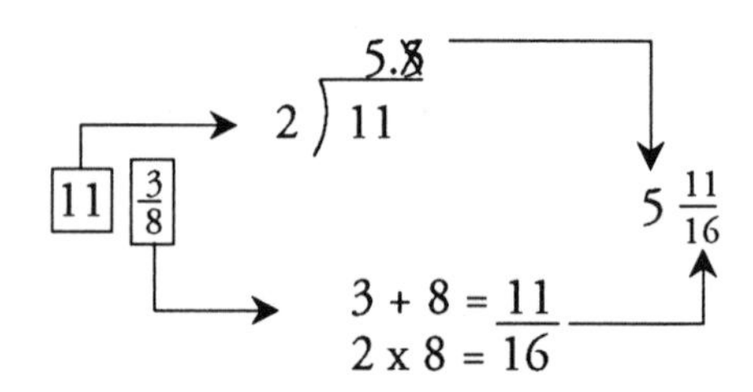

Rough Conversions

I also learned a trick from Jeff Bailey for making a rough conversion of decimals of a foot to inches and fractions of an inch. Each 0.01' is roughly equal to $\frac{1}{8}$". There are ninety-six $\frac{1}{8}$" in a foot and one hundred 0.01' in a foot. These facts comes in handy with grading when the surveyor tells you that you need to take off another 0.15'. A fast head calculation tells you that fifteen $\frac{1}{8}$" equals $\frac{15}{8}$ or $1\frac{7}{8}$". William Robertson told me the same trick except that for every three inches he adds the extra 0.01' to correct for the missing 0.04'.

The degree to which these tricks are effective depends on the degree of accuracy you need in your measurement.

Self Test

The key to using this book is to be aware of what you know and what you do not know. Just how much have you learned from your work in this book so far? Many people in school learn just enough to repeat the information back to the teacher or to pass a test. With this book, that is not enough. You may be seeing new material in the coming sections and you don't need to be struggling with converting fractions and decimals while learning something new. You need to know the previous sections backwards and forwards before continuing on. Below is a self test to help you determine whether you need to rework any of the previous sections or are ready to move on. The answers can be found on the next page.

Treat the numbers below either as measurements that need to be converted to decimals for calculator use or as answers from the calculator that need to be converted for application in the field. Convert to fractions or round off to four decimal places.

Remember: Look at the symbol for feet (') or inches (") to determine how to approach each problem.

(1) 11.6485"

(2) 5.9624'

(3) 42' 3"

(4) 14 $\frac{13}{16}$"

(5) 16.7437'

(6) 28' 5 $\frac{3}{16}$"

(7) 10.123"

(8) 33 $\frac{3}{4}$"

(9) 3.345'

(10) 148' 11 $\frac{5}{8}$"

(11) 93.7564'

(12) 23 $\frac{7}{8}$"

(13) 72.5649"

(14) 47' 6 $\frac{1}{16}$"

(15) 89.6789'

(16) 56.4569"

(17) 1.0032'

(18) 4 $\frac{3}{8}$"

(19) 24.9643"

(20) 47' 5 $\frac{7}{16}$"

If you worked through these problems with confidence, move on. If you felt a little awkward, work the needed sections again. In fact, throughout the book make it a habit to proceed only when you feel confident.

Answers to Self Test

(1)	$11\frac{5}{8}$"	(11)	93' $9\frac{1}{16}$"
(2)	5' $11\frac{9}{16}$"	(12)	23.875"
(3)	42.25'	(13)	$72\frac{9}{16}$"
(4)	14.8125"	(14)	47.5052'
(5)	16' $8\frac{15}{16}$"	(15)	89' $8\frac{1}{8}$"
(6)	28.4323'	(16)	$56\frac{7}{16}$"
(7)	$10\frac{1}{8}$"	(17)	1' $0\frac{1}{16}$"
(8)	33.75"	(18)	4.375"
(9)	3' $4\frac{1}{8}$"	(19)	$24\frac{15}{16}$"
(10)	148.9688'	(20)	47.4531'

III

Angles

Angle is an important term in the trades and has many definitions; however, the three definitions below cover most of the knowledge we need.

1. The shape made by two straight lines meeting in a point.

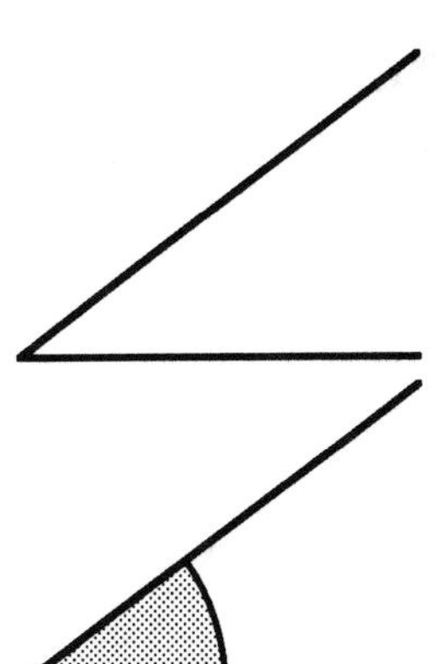

2. The space between those lines.

3. The amount of space measured in degrees.

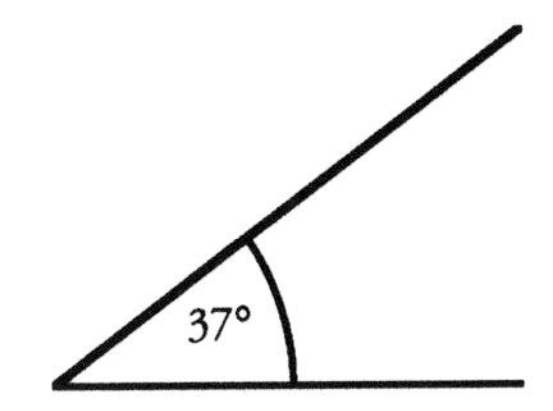

Vertex is the point where the two straight lines come together to form the angle.

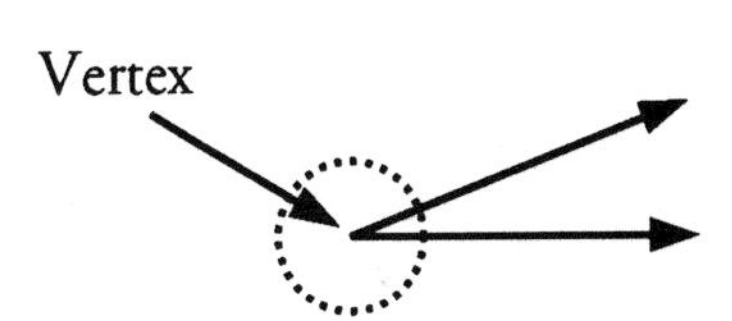

Angle Names

Names are given to angles with certain characteristics. Many of these names will be used in this book.

A **zero degree angle** has no space between the two lines.

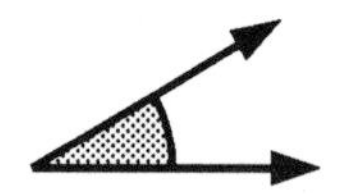

An **acute angle** is any angle between 0° and 90°.

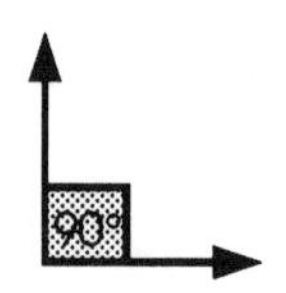

A **right angle** is a 90° angle. The two lines that form a 90° angle are **perpendicular** to each other. Notice that the angle symbol is in the shape of a box. That symbol indicates that you are dealing with a right angle.

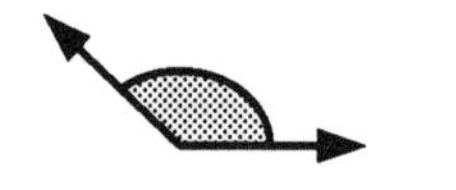

An **obtuse angle** is an angle between 90° and 180°.

A **straight angle** is an angle of 180°.

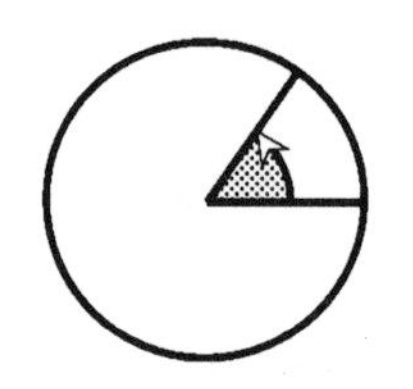

A **central angle** is an angle of which the vertex is located in the center of a circle.

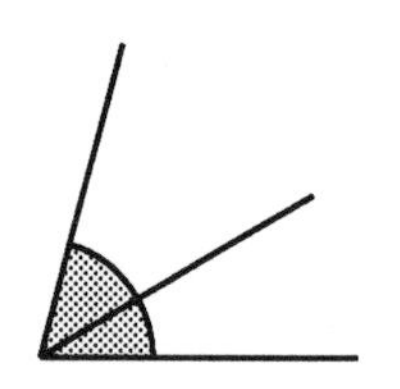

Adjacent angles are angles that have the same vertex and one side in common. (Adjacent means next to or near).

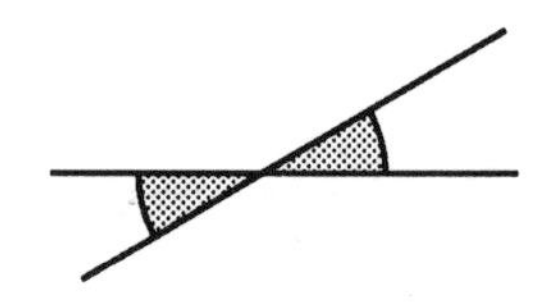

Opposite angles are formed by two straight lines crossing and are always equal.

Seeing is Believing-Page 185

90° Angles

The angle most often used in the trades is the 90° angle. It may help to understand why this is so. When two straight lines cross, four angles are created. Those four angles *always* add up to 360°.

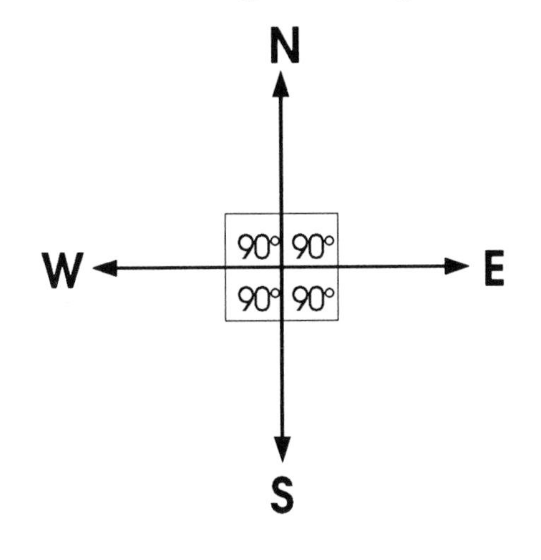

The major lines of the compass are two straight lines which cross each other at 90° angles. Most plants and factories are said to be built along the major lines of the compass.

When lines cross, the only time all four of the angles created are equal to each other is when each angle is a 90° angle. **A 90° angle is called a right angle**.

Seeing is Believing Page 183

Much of our work is lined up to be **level** or **plumb**.

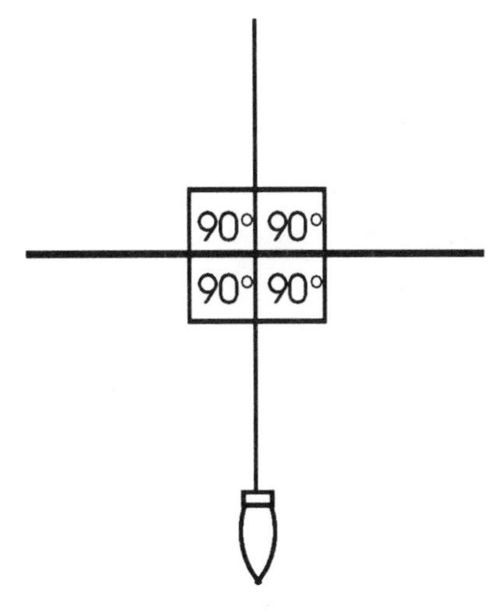

If you drop a plumb bob line across a level line, you again have two straight lines crossed at 90° angles.

Unless there are particular reasons for this not to be so, everything on a construction job is set up to be in line with a major compass line (axis) and to be plumb or level. That means that almost all things on a plant site are either parallel or perpendicular to each other. The easiest way to keep things that way is to make a majority of the turns at 90° angles.

Crossing and Parallel Lines

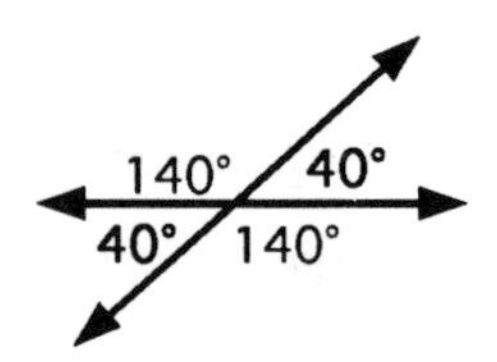

When two straight lines cross, four angles are formed.

Notice that the adjacent angles total 180° and the opposite angles are equal.

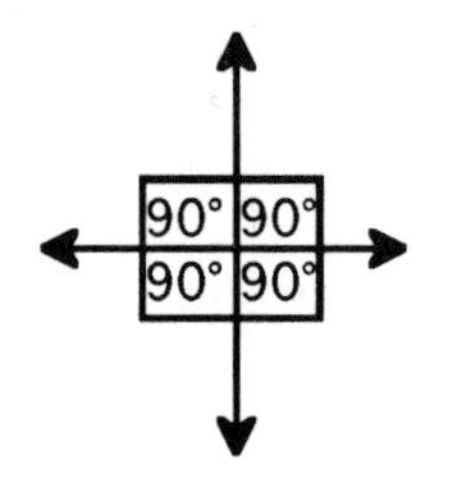

If two lines cross where all four angles are equal, the lines are **perpendicular** to each other.

Notice that the adjacent angles total 180° and the opposite angles are equal.

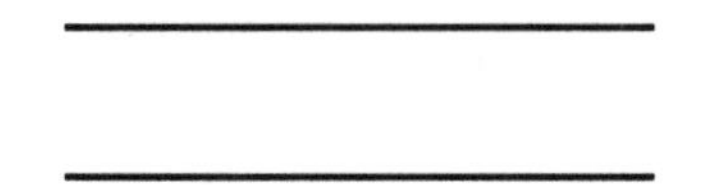

Two lines are **parallel** if they are *always* the same distance apart and in the same plane.

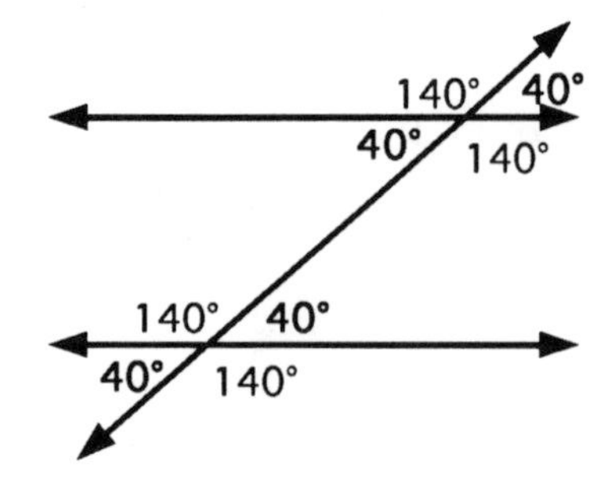

When a straight line crosses parallel lines, it crosses them at the same angles. *This is important in working with offsets, as you will see later.*

Notice that the adjacent angles total 180° and the opposite angles are equal.

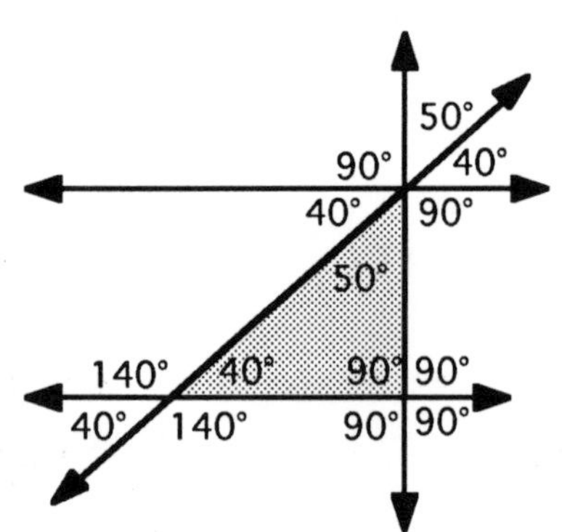

Here is an example illustrating all the above facts.

Notice that all the angles that form straight lines add up to 180°.

Notice that all opposite angles are equal.

Notice that a right triangle is formed by the crossing lines.

Notice that each line crosses the parallel lines at the same angle.

Parallel Lines

Parallel lines are lines in the same plane that never meet no matter how far they run. Two words in that statement that seem to confuse many people are *plane* and *parallel.* Let's look at these terms in a practical sense.

A plane is a flat surface that has length and width, but no thickness. Surface is defined as the exterior of an object. Surface has length and width. You can see it or imagine it, but because it has no thickness, you can't hold on to it.

Take for example the surface of a table top. You can see that it has length and width, you can even run your hand over it, but you can't pick up just the surface. It has no depth. That's the way a plane is. The difference between the table top surface and a plane is that the table top is not *perfectly* flat and a plane always is. The only time (in this world) we can have anything really perfect is in our imagination, and that's where a plane is. You have to imagine a perfectly flat surface that has length and width, but no thickness.

Of course, to explain or work with ideas using the term *plane*, we must imagine objects with similar characteristics as planes, even though they may not be perfect. For instance: a table top, a blackboard or this piece of paper may not be a "perfect" plane. Nevertheless, we can think of them as planes.

Parallel lines are *always* the same distance apart and always in the same plane.

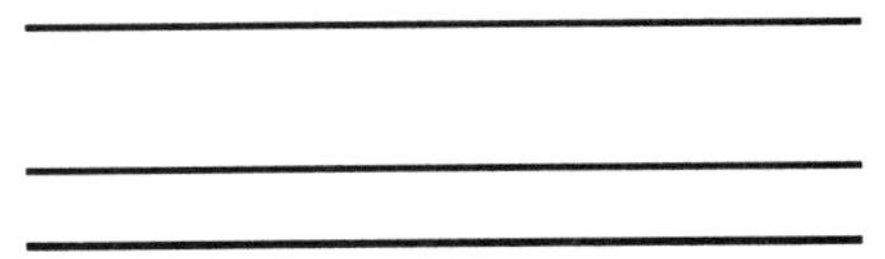

These lines are parallel to each other because they are on the same plane (the surface of this piece of paper) and they are the same distances apart.

Most materials used in the trades have opposite sides that are parallel. A board of lumber is usually rectangular or square, and each edge is parallel to the edge opposite it. The opposite sides of I beams are parallel. Paper has parallel sides.

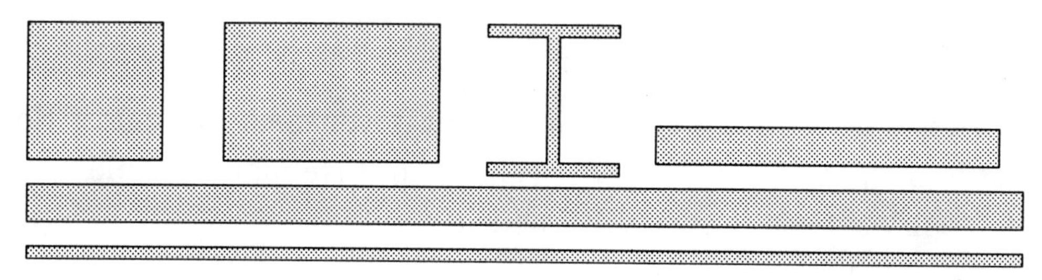

If you mark two straight lines down opposite sides of a pipe, you have drawn parallel lines.

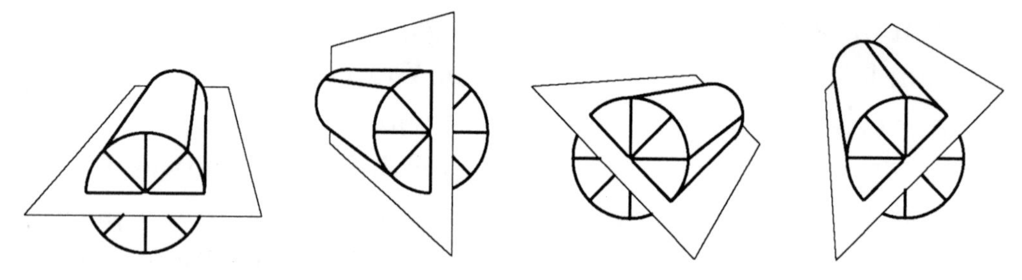

Note: This is not to say that you cannot draw lines that are not in the same plane and not parallel, but the lines we draw in our work usually are.

notes

IV

Degrees

***Degrees** are the units of measurement for angles.*

There are 360 degrees in any circle, and one degree is equal to $\frac{1}{360}$ of the complete rotation of a circle. 360 may seem to be an unusual number to use, but this part of math was developed in the ancient Middle East. During that era, the calendar was based on 360 days in a year, and one degree was equal to one day.

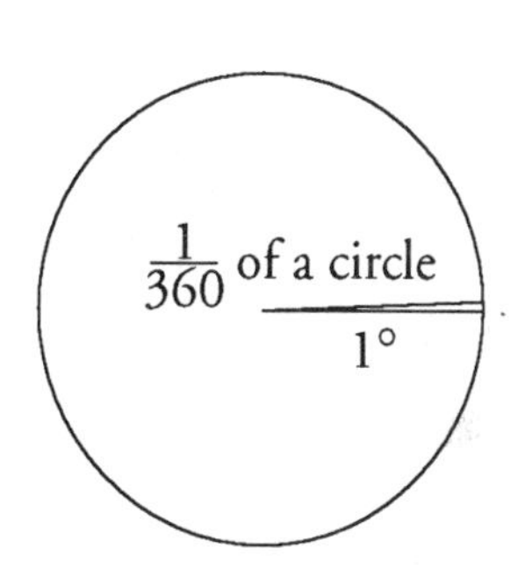

Fractions of Degrees

There are two methods of expressing fractions of degrees.

The first method divides each degree into 60 minutes (1° = 60'), then each minute into 60 seconds (1' = 60").

For example, you may see the degrees of an angle stated like this:

37° 42' 17" The symbol for degrees is °, for minutes is ', and for seconds is ".

The second method states the fraction as a decimal of a degree. This is the method we will use.

An example is 37° 42' 17" expressed as 37.7047°.

Most scientific calculators can display degrees both ways. The key for degrees on my calculator looks like [° ' "], but the key on another brand may look like [DMS]. You will need to refer to your calculator manual to determine the correct keys for degrees. Most calculators display answers in the form of degrees and a decimal of a degree.

It is seldom necessary to convert from minutes and seconds to decimals or vice versa; however, if you use the function tables of many trade manuals, it is necessary. Some tables show the fractions of degrees in minutes and seconds (DMS) rather than decimals (DD). In order to calculate using the different function tables, you must be able to convert the fractions to either format.

Converting Degrees, Minutes, and Seconds to Degrees and Decimals

To convert degrees, minutes, and seconds (DMS) to degrees and decimals of a degree (DD):

First: Convert the seconds to a fraction.
Since there are 60 seconds in each minute, 37° 42' 17" can be expressed as $37°42\frac{17}{60}'$. Convert to 37° 42.2833'.

Second: Convert the minutes to a fraction.
Since there are sixty minutes in each degree, 37° 42.2833' can be expressed as $37\frac{42.2833}{60}°$. Convert to 37.7047°

Calculator example: 37° 42' 17"

Enter	Display
Enter 17 [÷] 60 [=]	0.283333333
[+] 42	42.28333333
[÷] 60	0.704722222
[+] 37	37.70472222

Practice 16: Convert these DMS to the DD form. Round off to four decimal places. Answers-Page 218.

(1) 89° 11' 15"

(2) 12° 15' 0"

(3) 33° 30'

(4) 71° 0' 30"

(5) 42° 24' 53"

(6) 38° 42' 25"

(7) 29° 30' 30"

(8) 0° 49' 49"

Converting Degrees and Decimals to Degrees, Minutes, and Seconds

To convert degrees and decimals of degrees (DD) to degrees, minutes, and seconds (DMS), reverse the previous process.

First: Subtract the whole degrees. Convert the fraction to minutes. Multiply the decimal of a degree by 60 (the number of minutes in a degree). The whole number of the answer is the whole minutes.

Second: Subtract the whole minutes from the answer.

Third: Convert the decimal number remaining (from minutes) to seconds. Multiply the decimal by 60 (the number of seconds in a minute). The whole number of the answer is the whole seconds.

Fourth: If there is a decimal remaining, write that down as the decimal of a second.

Example: Convert 5.23456° to DMS.

5.23456° - 5° = 0.23456°	5° is the whole degrees
0.23456° x 60' per degree = 14.0736'	14 is the whole minutes
0.0736' x 60" per minute = 4.416"	4.416" is the seconds

DMS is stated as 5° 14' 4.416"

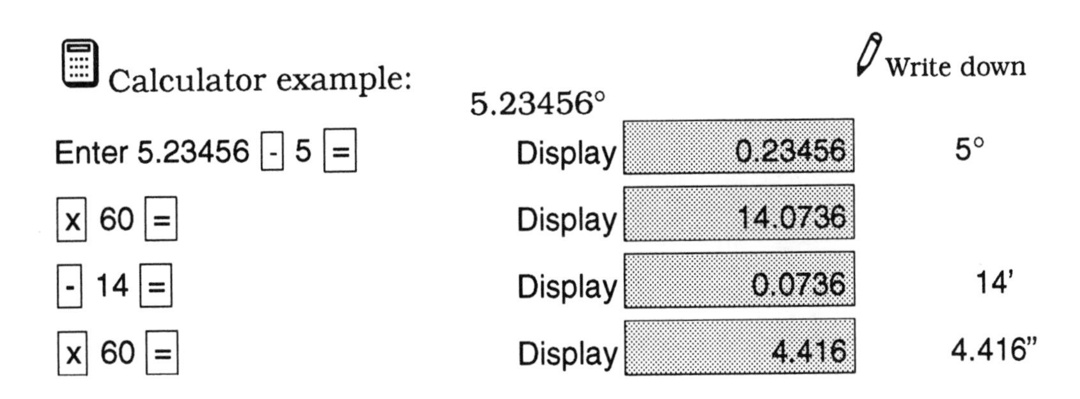

Practice 17: Convert these DD to DMS. Answers-Page 218.

(1) 75.25°

(2) 45.375°

(3) 9.5625°

(4) 33.9645°

(5) 13.12345°

(6) 21.5°

(7) 59.7892°

(8) 65.1836°

notes

Circles

You already know that **a circle has 360°**. Circles can be divided into different sectors: a half circle or semi-circle has 180°, a quarter circle has 90°, and an eighth of a circle has 45°.
This topic is covered in more detail in another section of this book.

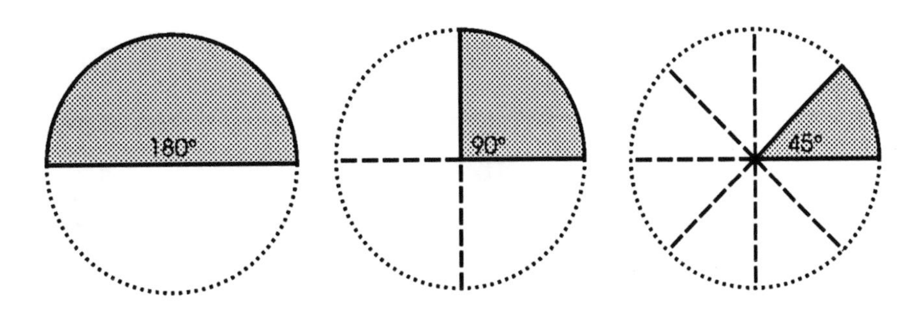

Triangles

A triangle is a closed figure that has three sides and three angles.
The sum of the three angles of every triangle *always* equals 180°.

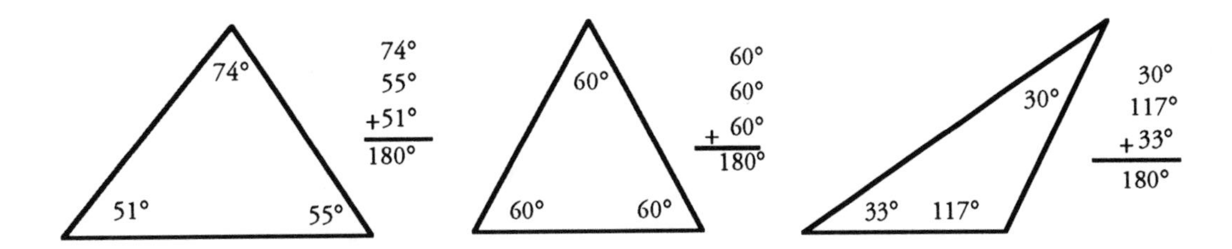

A right triangle *always* has one angle equal to 90° and two angles whose sum is 90°, making a total of 180°.

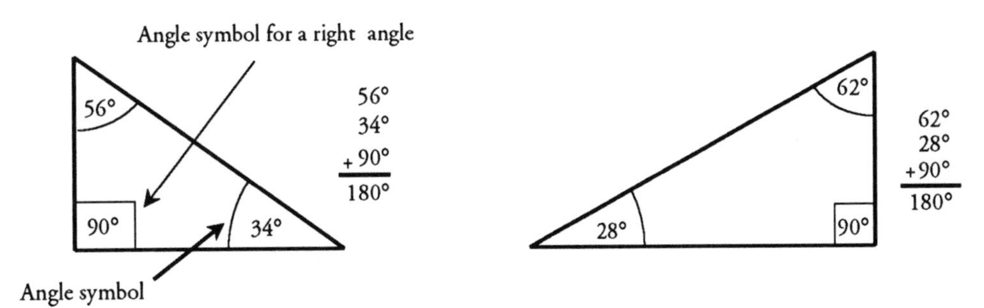

Rectangles

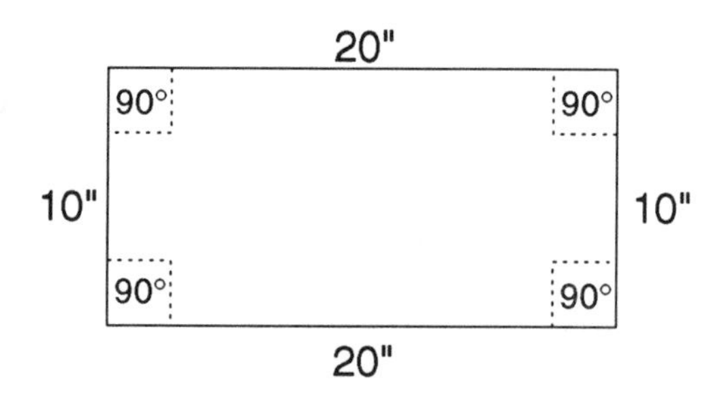

A rectangle is a four sided shape with four right angles (90°). *The sides across from each other are always the same length and always parallel to each other. The four angles add up to a total of 360°.*

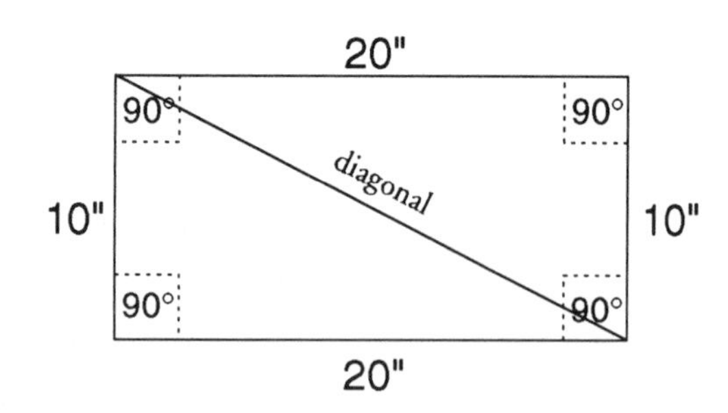

Another important rule about rectangles is that the **diagonals of a rectangle are equal.** Diagonals are lines that run between opposite corners. ***A diagonal always divides a rectangle exactly in half and creates two equal right triangles.***

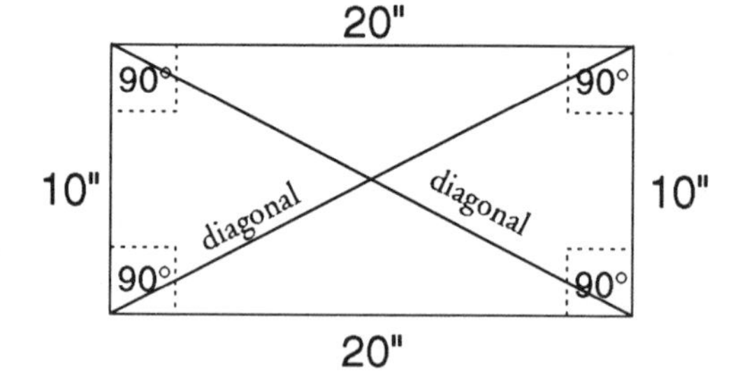

Carpenters doing layout work on forms measure both diagonals to make sure they are the same length. This insures that their layout is a rectangle which has square corners.

Squares

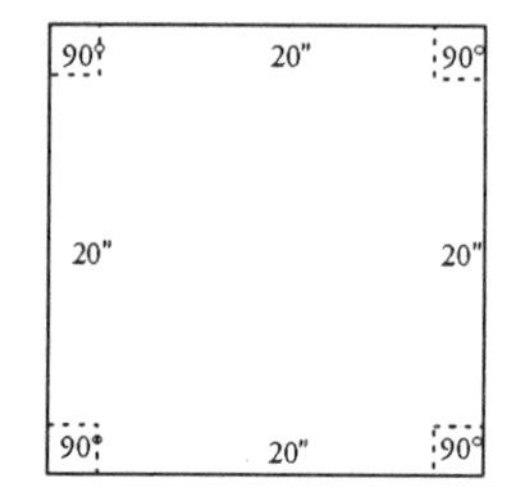

The square is a special type of rectangle in which all sides are equal. Since the square is a rectangle, the same rules that apply to the rectangle also apply to the square.

A diagonal across the square creates two equal 45° right triangles.

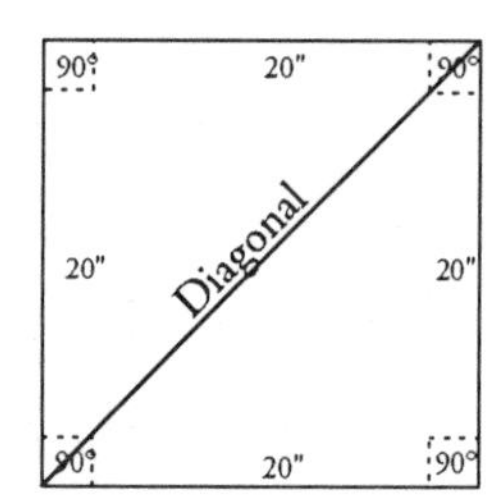

Later, you will be working with 45° right triangles.

It will be helpful to remember that ***the 45° right triangle is half of a square.***

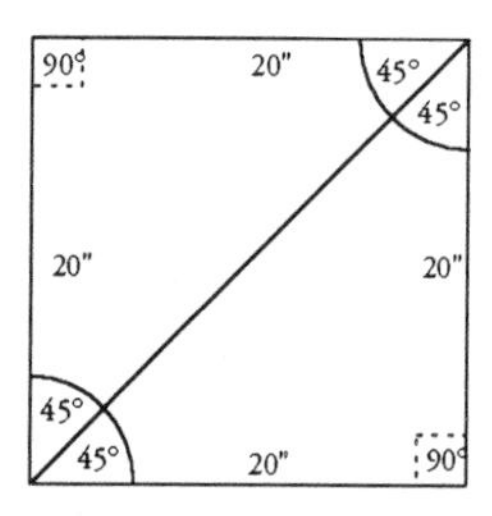

V

Variables and Constants

Variables

Some people get uncomfortable every time they see a letter of the alphabet mixed with numbers. They understand what to do with the numbers, but not those letters. The letters represent numbers that change in value or are unknown. They are called **variables** since the numbers those letters represent vary from time to time. A good example is pay. If a person is paid $20 per hour, to find his weekly pay, you would multiply $20 by the number of hours he works each week. The problem is that you may not know how many hours he will work each week. It may be 40 hours one week and 37 the next. However, if you represent the amount of hours with an **h**, you can say,

$20 x **h** = weekly pay.

The **h** is a variable that represents a real number that varies from week to week. The $20 represents the hourly rate of pay. To work with any variable, you must know exactly what it represents. There are many variables in this book, and, as you work with them, you will learn what they represent.

Constants

Constants are letters or symbols that represent one number. A well known constant is π (pi). Most people remember π as 3.14. You will learn more about π in the section on circles. *Many trades develop their own set of constants that work in certain situations.* Like shortcuts, they are fine when you know why you are using them.

notes

VI

Squaring Numbers

A number is squared when it is multiplied by itself. On the calculator, [x^2] will **square** whatever is in the display.

Example

Enter 3	Display	3
[x^2]	Display	9

Here are some examples of squares:

$$1^2 = 1 \times 1 = 1$$
$$2^2 = 2 \times 2 = 4$$
$$3^2 = 3 \times 3 = 9$$
$$4^2 = 4 \times 4 = 16$$
$$5^2 = 5 \times 5 = 25$$
$$6^2 = 6 \times 6 = 36$$
$$7^2 = 7 \times 7 = 49$$
$$8^2 = 8 \times 8 = 64$$
$$9^2 = 9 \times 9 = 81$$
$$10^2 = 10 \times 10 = 100$$
$$a^2 = a \times a = a^2$$
$$b^2 = b \times b = b^2$$
$$c^2 = c \times c = c^2$$

The variables in the last three examples represent any number or an unknown number. The variables show us the rule that any number squared is the number times itself.

Most of the numbers that you will square are mixed numbers, such as inches and fractions or feet, inches, and fractions. **Fractions and mixed numbers are easier to square if they are first converted to decimals.**

Example: What is the square of 4' $9\frac{7}{16}$"?

$$\boxed{(4'\ 9\tfrac{7}{16}")^2} = (4\ \frac{9\frac{7}{16}}{12}')^2 = (4\ \frac{9.4375}{12}')^2 = (4.786456333)^2 = 22.9102'$$

Enter 7 [÷] 16 [=] Display .4375

[+] 9 [=] Display 9.4375

[÷] 12 [=] Display .786458333

[+] 4 [=] Display 4.786458333

[x^2] Display 22.91018338

Practice 18: Use the [x^2] key to find these squares. Round off to 4 decimal places. Answers-Page 218.

(1) 1^2

(2) 10^2

(3) 130^2

(4) 7.5^2

(5) 5.525^2

(6) $8\frac{3}{8}^2$

(7) 11^2

(8) d^2

VII

Square Roots

The **square root** of a number is best described by the question: What number has been squared to get the number under the square root symbol? The square root of 4, ($\sqrt{4}$) is 2. That means that if 2 is squared, it will equal 4. The $\boxed{\sqrt{x}}$ key will find the square root of whatever is in the display on your calculator.

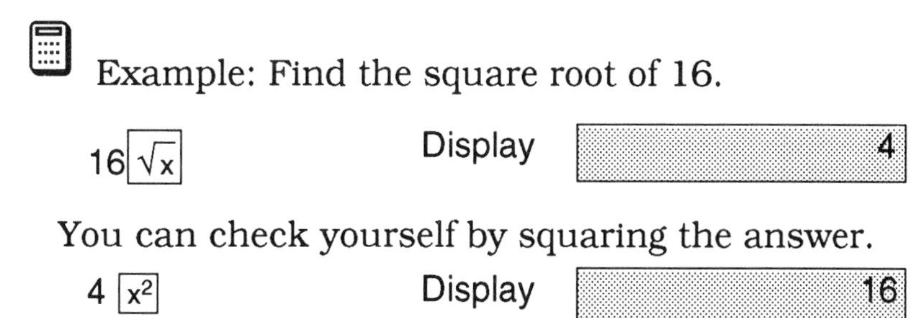

Here are some examples of square roots.

$\sqrt{1} = 1$

$\sqrt{4} = 2$

$\sqrt{9} = 3$

$\sqrt{16} = 4$

$\sqrt{25} = 5$

$\sqrt{36} = 6$

$\sqrt{49} = 7$

$\sqrt{a^2} = a$

$\sqrt{b^2} = b$

$\sqrt{c^2} = c$

Every number has a square and a square root.

The variables in the last three examples show another rule of math: *The square root of a number squared is the number.* For example: $\sqrt{4}$ or $\sqrt{2^2} = 2$.

Practice 19: Find the square root for each of the following numbers. Use the $\boxed{\sqrt{x}}$ key. Round off to 4 decimal places. Answers-Page 218.

(1) 1

(2) 22

(3) 54.5

(4) $16\frac{3}{4}$

(5) 400

(6) 127

(7) 74.525

(8) $35\frac{7}{8}$

(9) a^2

(10) b^2

(11) 144

(12) 100

VIII

Right Triangles

The right triangle has been used in trades for thousands of years. Ancient Egyptians found that they could always get a square corner using the 3-4-5 right triangle. (Carpenters still use the 3-4-5 triangle to square corners.) Later, a Greek mathematician named Pythagoras developed a formula to find the side lengths for any right triangles.

Pythagoras treated each side of a right triangle as if it was a side of a square. He found that the total area of the two smaller squares was equal to the area of the largest square for every right triangle.

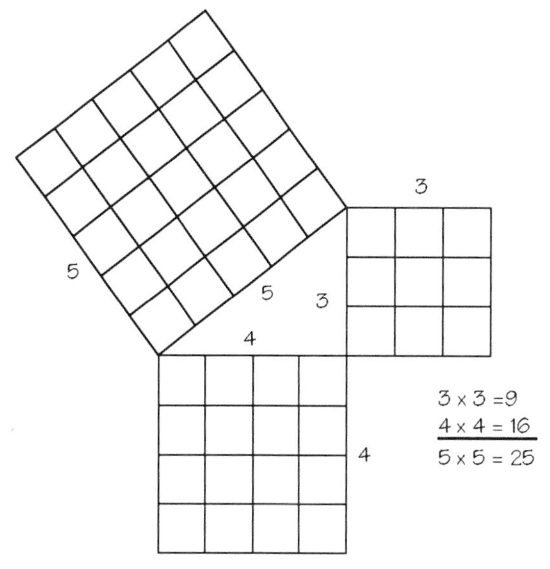

Take a look at the diagram to the right. The area of any square is found by squaring one of the sides. The areas of these squares are 3^2, 4^2, and 5^2. If the two shorter sides are squared and added together, the answer equals the longer side squared: $9 + 16 = 25$

$5^2 = 25$

The three sides of a **right triangle** are represented by the variables **a**, **b**, and **c**. The variable **c** *always* represents the longest side. You can write the formula used above as $a^2 + b^2 = c^2$.

This is the formula of the Pythagorean Theorem.

$$\mathbf{a^2 + b^2 = c^2}$$

The longest side of a right triangle is *always* directly across from the 90 degree angle. This side is called the **hypotenuse**.

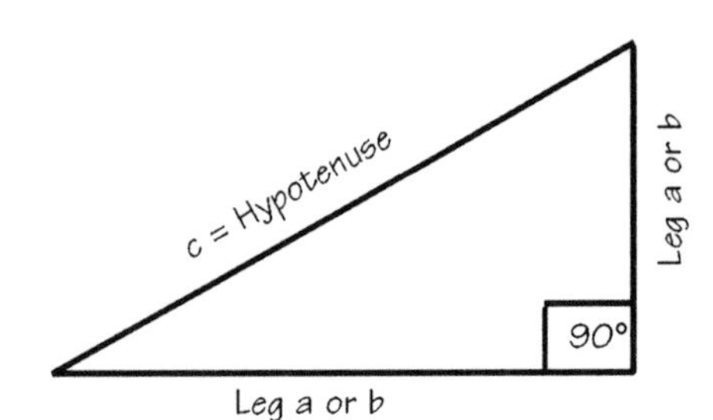

The other two sides are often called the **legs**. For now, it really doesn't matter which one of the legs you call **a** or **b**, as long as you make one **a** and the other **b**.

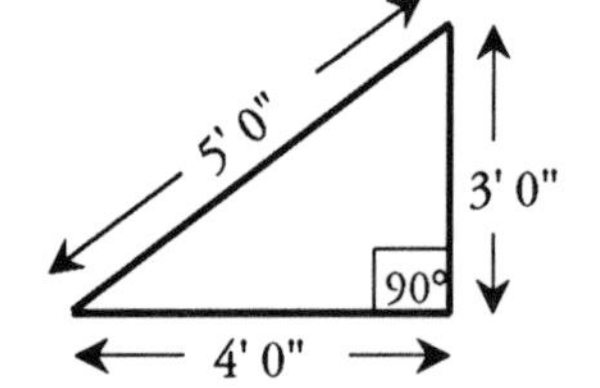

Because the 3-4-5 right triangle is simple to work with, let's use it to show how the Pythagorean formula *verifies that a triangle is a right triangle.*

a = 3 b = 4 c = 5

$$a^2 + b^2 = c^2$$

$$3^2 + 4^2 = 5^2$$

$$9 + 16 = 25$$

$$25 = 25$$

Since the numbers on both sides of the = mark are the same, the 3-4-5 triangle is a right triangle.

Please note: The Pythagorean formula is a rule! To verify that a triangle is a right triangle, use the $a^2 + b^2 = c^2$ formula. If the numbers on each side of the = sign are equal, then the triangle is a right triangle.

Practice 20: Check these triangles to see which ones are right triangles. Answers- Page 218.

	a	b	c
(1)	6	8	10
(2)	.48	.64	.8
(3)	1	1	1
(4)	8	6	10
(5)	120	160	200

	a	b	c
(6)	9.99	13.32	16.65
(7)	4.5	6	7.5
(8)	7	10	13
(9)	9	2.5	13
(10)	2	2.6666	3.3333

Note: Lower case letters (a,b,c) are used as variables for the sides. Upper case letters (A,B,C) are used for the angles opposite those sides.

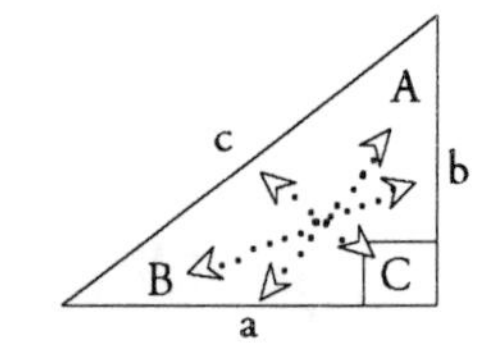

Finding the Length of the Hypotenuse

The formula $a^2 + b^2 = c^2$ can be changed around to find the length of an unknown side when the other two sides are known.

The formula $\sqrt{a^2 + b^2} = c$ ***is used to find the length of the hypotenuse when the lengths of the two legs are known.***

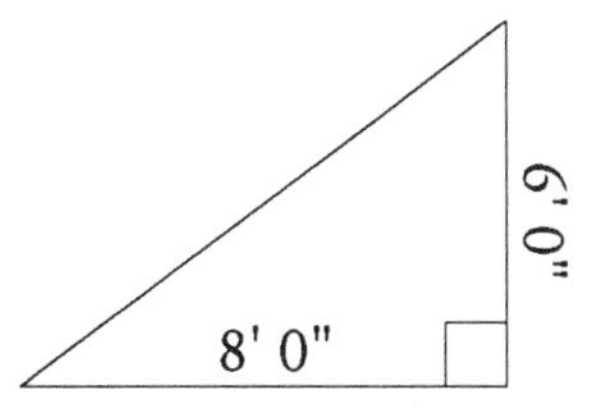

In this example, we know the length of the legs *a* and *b*, but not the length of the hypotenuse, *c*.

To find the length of the hypotenuse,

- *Replace the variables of the formula with the lengths of the legs.*
- *Do the calculations.* The answer is the length of the hypotenuse c.

$$\sqrt{a^2 + b^2} = c$$
$$\sqrt{6^2 + 8^2} = c$$
$$\sqrt{36 + 64} = c$$
$$\sqrt{100} = c$$
$$\mathbf{10} = c$$

STOP Warning

Do not add $6^2 + 8^2$ to get 14^2. It doesn't work!

Here is this formula worked on the calculator.

Remember: The keys $\boxed{x^2}$ and $\boxed{\sqrt{x}}$ react to whatever is on the display when they are pushed.

Keys	Display
Enter 6 $\boxed{x^2}$ $\boxed{+}$	36
Enter 8 $\boxed{x^2}$	64
$\boxed{=}$	100
$\boxed{\sqrt{x}}$	10

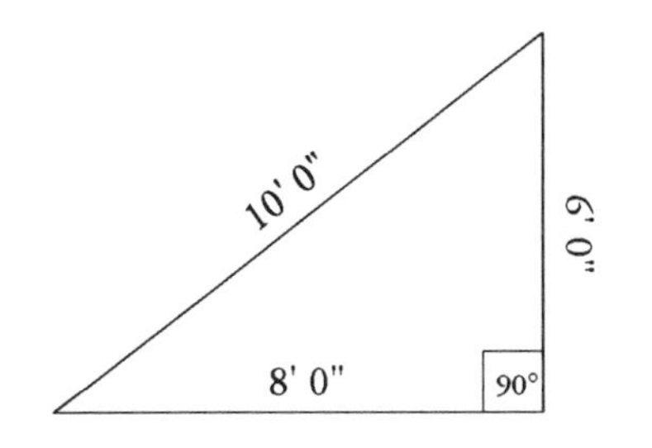

10' is the length of the hypotenuse.

Practice 21: Find the length of the hypotenuse using the two sides given. Convert the numbers with units of measurement to include fractions. Round off the rest of the numbers to 4 decimal places. Answers-Page 218.

If you have difficulty squaring mixed numbers, refresh your memory on page 43.

c, a, b or c, b, a

	a	b		a	b
(1)	6	9	(7)	9' 4 $\frac{3}{4}$"	12' 11"
(2)	5	10	(8)	1.73"	2.32"
(3)	22"	33"	(9)	104' 3"	77' 7"
(4)	2.5"	7.5"	(10)	19	25
(5)	10 $\frac{3}{8}$"	7 $\frac{7}{16}$"	(11)	16 $\frac{15}{16}$"	22 $\frac{15}{16}$"
(6)	7' 8"	3' 6"	(12)	24	44

notes

Finding the Length of a Leg

If you know the lengths of the hypotenuse and one leg and need to know the length of the other leg, the formula for the right triangle can again be rearranged to find the answer. The variable that represents the unknown side will need to be on a side of the formula by itself. Since you are naming the sides, the variable can be either *a* or *b*.

The formula needed to find an unknown leg when the other sides are known is:

$$\mathbf{a = \sqrt{c^2 - b^2}} \text{ or } \mathbf{b = \sqrt{c^2 - a^2}}$$

Notice, as indicated under the square root symbol, that you subtract the square of the length of the known leg from the square of the length of the hypotenuse to find the unknown leg.

Remember: The hypotenuse is always the longest side and therefore the largest number. Because of this, the leg^2 is always subtracted from the $hypotenuse^2$.

Example: If the length of the hypotenuse is 150 and one of the legs is 90, what is the length of the other leg? Let's say 90 is side b.

Remember: The hypotenuse is always c.

When we put the above numbers into the formula, the calculations look like this.

$a = \sqrt{c^2 - b^2}$

$a = \sqrt{150^2 - 90^2}$

$a = \sqrt{22500 - 8100}$

$a = \sqrt{14400}$

$a = \mathbf{120}$

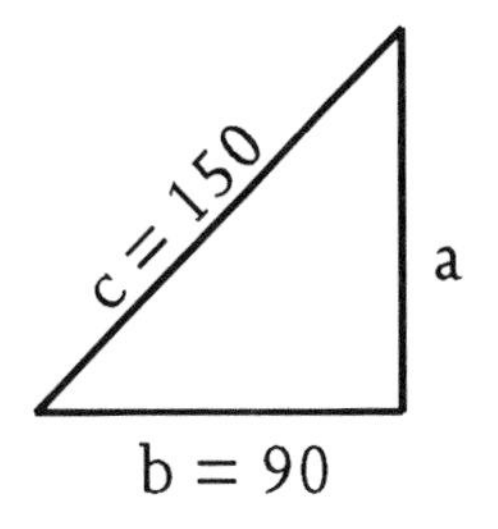

Here is the formula worked using the calculator.

Enter 150 [x^2] [-]	Display 22500	
Enter 90 [x^2]	Display 8100	
[=]	Display 14400	
[$\sqrt{x}$]	Display 120	120 is the third side, a.

Practice 22: Find the length of the third side. Convert the numbers with units of measurement to a fraction format. Round the rest of the numbers off to 4 decimal places. Answers-Page 218.

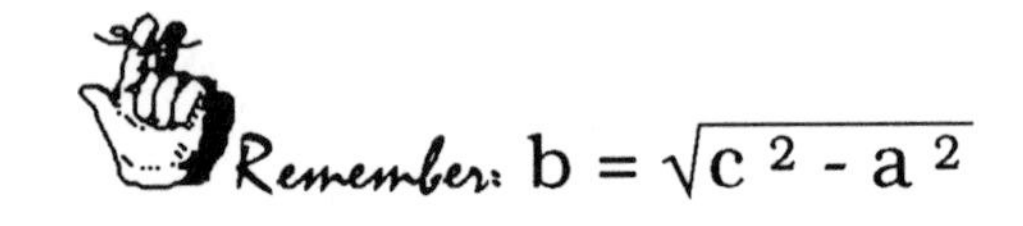

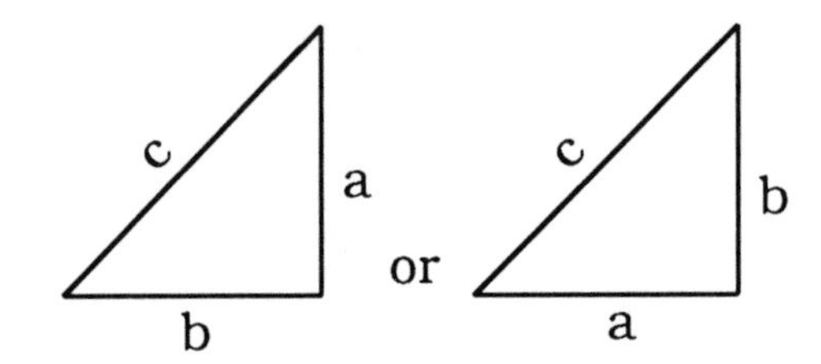

	a	c		a	c
(1)	15	20	(7)	8.375	13.5625
(2)	$6\frac{1}{2}$"	$10\frac{7}{16}$"	(8)	32.975 '	52.25'
(3)	12	18	(9)	4	16
(4)	6'4"	9'5"	(10)	$12\frac{7}{8}$"	13"
(5)	88'	100'	(11)	5'	10'
(6)	18"	22"	(12)	1' $0\frac{1}{4}$"	2'

Names of the Sides of a Right Triangle

Knowing the names of the sides of a right triangle is essential. Besides being a part of the vocabulary of math, these names are used to communicate which side of a triangle you are referring to without using a diagram. Also if you misname the sides, your calculations of ratios (which you are about to study) will be off.

The **hypotenuse** is the side directly across from the 90° angle and is always the longest side of a right triangle.

The other two sides are named according their position to the **reference angle,** which is the angle you are referring to or seeking. **The reference angle is *never* the 90° angle**.

The **adjacent side** is the side next to the reference angle.

The **opposite side** is the side directly across from the reference angle.

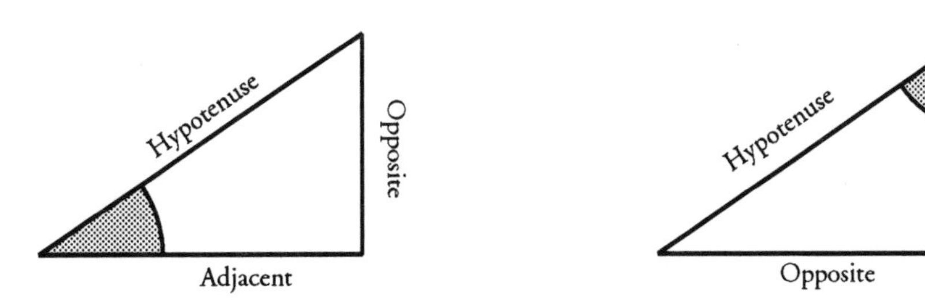

The gray shading indicates the reference angle.

*It is important to understand that the names of the opposite side and adjacent sides **change** when you move from one reference angle to the other.*

Ratio of Sides

A ratio is a comparison by division. You divide one factor into another to see how they compare.

In sports, ratios are constantly flashed on the TV screen. With baseball, we are shown a player's batting average, which compares his number of hits to his number of times at bat.

$$\text{Batting Average} = \frac{\text{Hits}}{\text{At Bat}}$$

In basketball, a ratio of shots made to shots taken is called shooting average.

$$\text{Shooting Average} = \frac{\text{Shots made}}{\text{Shots taken}}$$

The ratio of sides of a right triangle is determined by comparing the lengths of two sides.

Example: $\frac{\text{opposite side}}{\text{adjacent side}}$ or $\frac{\text{hypotenuse}}{\text{opposite side}}$

To introduce ratios of the sides, let's use the 3-4-5 right triangle again.

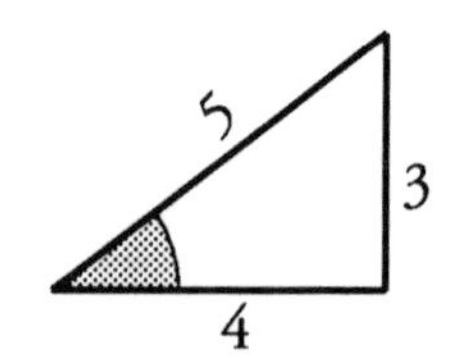

With each angle, there are **only** six possible ways that the sides can be divided into each other, so there are **only** six possible ratios for each angle. The six ratios for the reference angle shown (indicated by the shading) are as listed:

Hypotenuse
Opposite
Adjacent

$$\frac{\text{Opposite side}}{\text{Hypotenuse}} = \frac{3}{5} \qquad \frac{\text{Hypotenuse}}{\text{Opposite side}} = \frac{5}{3}$$

$$\frac{\text{Adjacent side}}{\text{Hypotenuse}} = \frac{4}{5} \qquad \frac{\text{Hypotenuse}}{\text{Adjacent side}} = \frac{5}{4}$$

$$\frac{\text{Opposite side}}{\text{Adjacent side}} = \frac{3}{4} \qquad \frac{\text{Adjacent side}}{\text{Opposite side}} = \frac{4}{3}$$

Each ratio has been assigned a name, and these names are called **functions**.

This chart shows the functions and their ratios.

Sine θ	=	$\frac{\text{Opposite side}}{\text{Hypotenuse}}$	Cosecant θ	=	$\frac{\text{Hypotenuse}}{\text{Opposite side}}$
Cosine θ	=	$\frac{\text{Adjacent side}}{\text{Hypotenuse}}$	Secant θ	=	$\frac{\text{Hypotenuse}}{\text{Adjacent side}}$
Tangent θ	=	$\frac{\text{Opposite side}}{\text{Adjacent side}}$	Cotangent θ	=	$\frac{\text{Adjacent side}}{\text{Opposite side}}$

θ is the Greek letter **theta**. It is a variable which is used to represent the degrees of an angle when the degrees are unknown. In the chart above, θ is used to represent any angle. For example, the sine function can be read: The sine of *any angle* is the length of the hypotenuse divided into the length of the opposite side.

This chart shows the functions and the ratios for the reference angle. Notice that the ratios are expressed as both fractions and decimals.

Sine θ	$\frac{O}{H}$	$\frac{3}{5}$	0.6000
Cosine θ	$\frac{A}{H}$	$\frac{4}{5}$	0.8000
Tangent θ	$\frac{O}{A}$	$\frac{3}{4}$	0.7500

Cosecant θ	$\frac{H}{O}$	$\frac{5}{3}$	1.6666
Secant θ	$\frac{H}{A}$	$\frac{5}{4}$	1.2500
Cotangent θ	$\frac{A}{O}$	$\frac{4}{3}$	1.3333

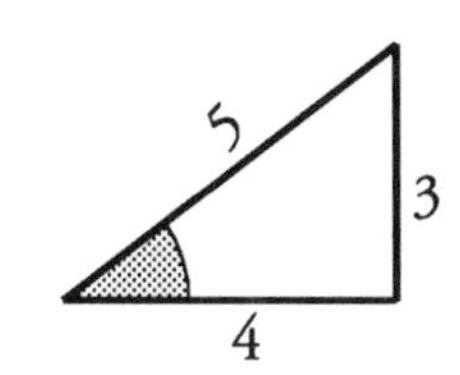

The hypotenuse is always the longest side. Notice that when the length of the hypotenuse is the numerator, the ratio is always greater than one. When the length of the hypotenuse is the denominator, the ratio is always less than one.

Here are the standard abbreviations for the functions.

Sin θ	=	Sine θ	Csc θ	=	Cosecant θ
Cos θ	=	Cosine θ	Sec θ	=	Secant θ
Tan θ	=	Tangent θ	Cot θ	=	Cotangent θ

Practice 23: Determine the six functions for each angle of these right triangles. Round off to 4 decimal places. Answers-Page 218.

Remember: There are two reference angles in each right triangle. The right angle is never the reference angle.

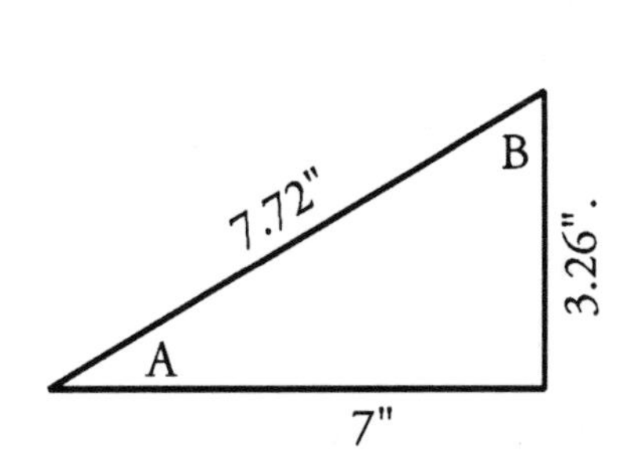

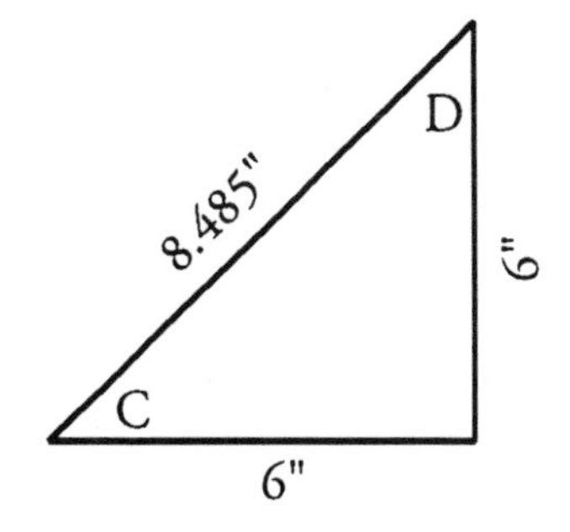

Memory Aid

There is a learning trick that can help you keep the functions straight. Simply remember this sentence: **Oscar Had A Heap Of Apples.** Memorizing this sentence allows you to keep the functions in correct order. As you can see, this phrase uses the first letter of each of the sides.

Oscar	**H**ad	**A**	**H**eap	**O**f	**A**pples
Opposite	**H**ypotenuse	**A**djacent	**H**ypotenuse	**O**pposite	**A**djacent

Sin $= \dfrac{\textbf{Opposite side}}{\text{Hypotenuse}}$ $\dfrac{\text{Oscar}}{\text{Had}}$

Cos $= \dfrac{\textbf{Adjacent side}}{\text{Hypotenuse}}$ $\dfrac{\text{A}}{\text{Heap}}$

Tan $= \dfrac{\textbf{Opposite side}}{\textbf{Adjacent side}}$ $\dfrac{\text{Of}}{\text{Apples}}$

What you have to remember is **sin, cos, tan,** and **Oscar had a heap of apples.**

The other functions are remembered by the order in which they are placed.

Sine	Cosecant	You can't have the two S's or the two Co's on the same line.
Cosine	Secant	You can't have the two S's or the two Co's on the same line.
Tangent	Cotangent	The tangent and the cotangent are easy to remember.

The third part of remembering the order is to realize that the functions on the right side of the chart are inverse to the functions on the left side. What this means is that the numerators and denominators are in opposite positions.

Sin $= \frac{O}{H}$	Csc $= \frac{H}{O}$
Cos $= \frac{A}{H}$	Sec $= \frac{H}{A}$
Tan $= \frac{O}{A}$	Cot $= \frac{A}{O}$

If you need the secant to an angle, just remember that the secant is across from the cosine (therefore its inverse) and cosine relates to the second pair of words in *Oscar Had **A Heap** Of Apples.*

This means that cosine is $\dfrac{\text{Adjacent side}}{\text{Hypotenuse}}$, so secant is $\dfrac{\text{Hypotenuse}}{\text{Adjacent side}}$.

Finding the Angles of a Right Triangle

A key to mastering the right triangles is understanding how the ratios of sides, the functions, and the angles are related. **The ratios of sides are directly related to the number of degrees in the reference angle.** A relationship between the ratios of sides and the number of degrees in the angles of a right triangle has been known for years. It was found that *a right triangle that has the same angles as another right triangle also has the same ratio of sides.* A system was worked out which enables us to compare the ratios to a chart (these days we use a calculator) and find the degrees of the angles. What this means is: ***If you take the time to learn the names of the functions and the ratios that are assigned to them, you will be able to find the degrees of the angles of any right triangle just by knowing the length of two sides.***

Using the Arc Function Keys on the Calculator

The scientific calculator has a group of keys called the **arc function** keys. The arc function symbol is the $^{-1}$ in the upper right corner above the function name. ***The arc function keys display the angle when given the correct function***.

There are three arc function keys on most calculators.

Sin^{-1}	**Arc sine**	This key when pushed shows the angle of the **sine** in the display of the calculator.
Cos^{-1}	**Arc cosine**	This key shows the angle of the **cosine** in the display.
Tan^{-1}	**Arc tangent**	This key shows the angle of the **tangent** in the display.

The three other arc functions — **arc cosecant**, **arc secant**, and **arc cotangent** — are also used to find angles, but require another step. For the present, use the three arc function keys on the calculator.

$\text{Sin}\ \theta =$	$\frac{\text{Opp}}{\text{Hypo}}$
$\text{Cos}\ \theta =$	$\frac{\text{Adj}}{\text{Hypo}}$
$\text{Tan}\ \theta =$	$\frac{\text{Opp}}{\text{Adj}}$

These three functions deal with all three sides of a right triangle, and ***Oscar Had A Heap Of Apples*** will help you remember these functions.

Example: Find the degrees of the reference angle of this 3-4-5 right triangle.

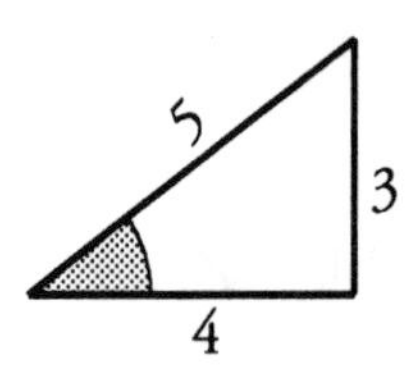

First: *Select the reference angle and name the legs.*

Remember: **The right angle is never used as the reference angle.**

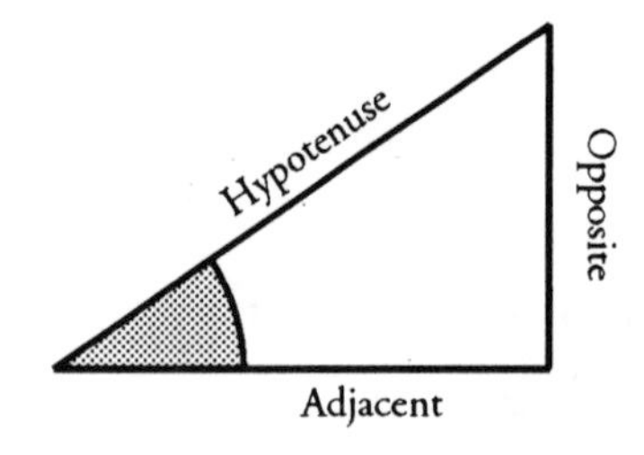

Second: *Look at the functions chart and choose the function that corresponds to the known sides of the triangle.*

- Since we know the lengths of all the sides of the above triangle, any of the functions can be used. Let's use the **sine** function, $\frac{O}{H}$.

Third: *Assign the numbers to the ratio and calculate the problem.*

- Sine $\theta = \frac{O}{H} = \frac{3}{5} = 0.6$

Fourth: *Use the correct arc function key to find the degrees of the reference angle.*

- The [sin⁻¹] key is used to find the degrees of this reference angle.

Example

Enter 3 [÷] 5 [=]	Display 0.6
Push [$\sin^{-1}$]	Display 36.86989765

Round off to 37° (nearest $\frac{1}{2}$ °)*

---

* In rounding off degrees to the nearest $\frac{1}{2}$ degree, from 0.00 to 0.249, go down to 0. From 0.25 to .749, go to .5°. From .75 to .99, go to 1°.

As you can see, the arc function keys react to whatever is in the display. On the next practice, enter the number, look at the function to determine which arc function key to use, and push that key.

Practice 24: Find the angle for each function. Round off to the nearest $\frac{1}{2}°$. Answers-Page 218.

(1) $\sin\theta = 0.3745$

(2) $\tan\theta = 1$

(3) $\cos\theta = 0.8632$

(4) $\sin\theta = 1$

(5) $\tan\theta = 0.5785$

(6) $\sin\theta = 0.7756$

(7) $\cos\theta = 0.9397$

(8) $\tan\theta = 1.5030$

notes

Steps to Finding Angles When Two Sides Are Known

Example: Use the two known side lengths of the right triangle below to find the angles of the triangle.

Remember: **The function table can be used to find an angle when you know two side lengths of a right triangle.**

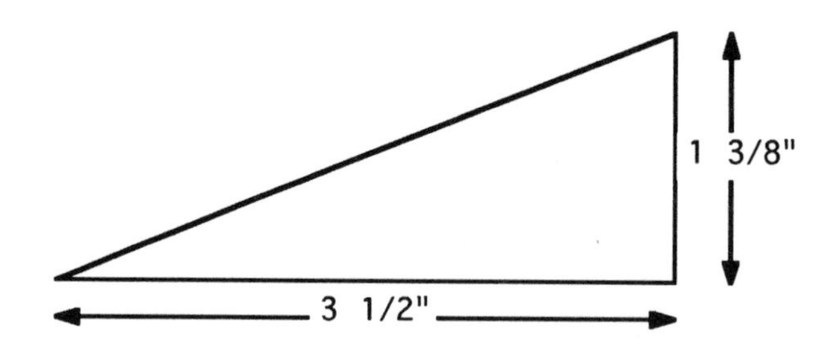

Follow the steps below.

First: Mark the reference angle and name the sides for that angle.

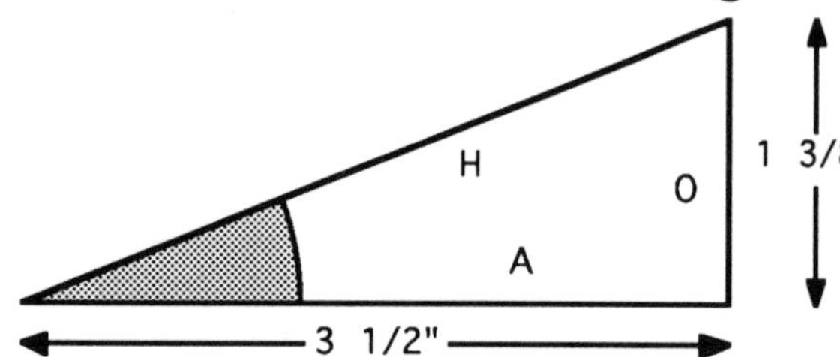

- Adjacent side = $3\frac{1}{2}$"
- Opposite side = $1\frac{3}{8}$"

Second: Look at the three functions and decide which one will work with the known sides. Replace the ratio with the leg lengths of the triangle and calculate the problem.

Sine θ	=	$\frac{\text{Opposite side}}{\text{Hypotenuse}}$
Cosine θ	=	$\frac{\text{Adjacent side}}{\text{Hypotenuse}}$
Tangent θ	=	$\frac{\text{Opposite side}}{\text{Adjacent side}}$

- Tangent is the correct choice. It is the only function of the three which uses both known sides.

$$\text{Tangent } \theta = \frac{\text{opposite side}}{\text{adjacent side}} = \frac{1.375}{3.5} = 0.392857142$$

Third: Use the correct arc function key to find the degrees of the reference angle.

- With the answer (0.392857142) still in the display, push the $\boxed{\tan^{-1}}$ key. The display will show 21.44773633, which is the degrees of the reference angle. If we round the number off to the nearest $\frac{1}{2}$ degree, it becomes **21.5°**.

Fourth: Find the degrees of the other angle.

Remember: **The sum of the three angles of any triangle will *always* equal 180°.**

- Since the two angles (other than the right angle) equal a total of 90°, subtract the known angle from 90° to find the degrees of the other angle.

  90° - 21.5° = **68.5°**

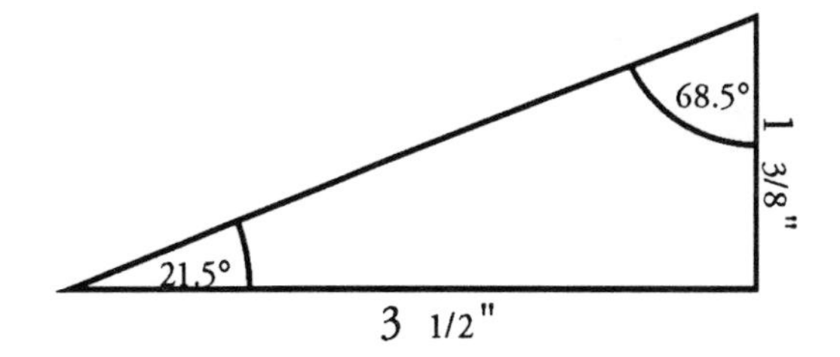

Practice 25: Find the measurement for both angles of each right triangle. Round off each answer to the nearest $\frac{1}{2}$°. Answers-Page 219.

(1)	Opp =	22"	Hyp =	34"
(2)	Adj =	19.375"	Opp =	27.5"
(3)	Adj =	64.9375"	Hyp =	78.4375"
(4)	Opp =	14.5"	Adj =	29.685"
(5)	Hyp =	3'	Opp =	2.5'

notes

Finding the Angle Using the Functions Table

This is a sample of the functions table on pages 214 and 215. Notice that the table includes all six functions, plus something you will study later, radians.

Deg↓	Radian↓	Sin θ↓	Cos θ↓	Tan θ↓	Cot θ↓	Sec θ ↓	Csc θ↓		
36.5°	0.6370	0.5948	0.8039	0.7400	1.3514	1.2440	1.6812	0.9338	53.5°
37°	0.6458	0.6018	0.7986	0.7536	1.3270	1.2521	1.6616	0.9250	53°
37.5°	0.6545	0.6088	0.7934	0.7673	1.3032	1.2605	1.6427	0.9163	52.5°
		Cos θ↑	Sin θ↑	Cot θ↑	Tan θ↑	Csc θ↑	Sec θ↑	Radian↑	Deg↑

The functions table is read in two directions:

For 0° to 45°, read top to bottom

Deg↓	Radian↓	Sin θ↓	Cos θ↓	Tan θ↓	Cot θ↓	Sec θ↓	Csc θ↓		
36.5°	0.6370	0.5948	0.8039	0.7400	1.3514	1.2440	1.6812	0.9338	53.5°
37°	0.6458	0.6018	0.7986	0.7536	1.3270	1.2521	1.6616	0.9250	53°
37.5°	0.6545	0.6088	0.7934	0.7673	1.3032	1.2605	1.6427	0.9163	52.5°
		Cos θ↑	Sin θ↑	Cot θ↑	Tan θ↑	Csc θ↑	Sec θ↑	Radian↑	Deg↑

For 45° to 90°, read bottom to top.

Deg↓	Radian↓	Sin θ↓	Cos θ↓	Tan θ↓	Cot θ↓	Sec θ↓	Csc θ↓		
36.5°	0.6370	0.5948	0.8039	0.7400	1.3514	1.2440	1.6812	0.9338	53.5°
37°	0.6458	0.6018	0.7986	0.7536	1.3270	1.2521	1.6616	0.9250	53°
37.5°	0.6545	0.6088	0.7934	0.7673	1.3032	1.2605	1.6427	0.9163	52.5°
		Cos θ↑	Sin θ↑	Cot θ↑	Tan θ↑	Csc θ↑	Sec θ↑	Radian↑	Deg↑

Notice that angles are shown on both sides of the chart. When reading *down* the column for 0° through 45°, use the function names listed at the top of the table. When reading *up* the table for 45° through 90°, use the function names at the bottom. *It is important that you note the difference.*

We worked with the 3-4-5 right triangle earlier. Let's use the same angle for an example again. Below are the functions for the reference angle of that 3-4-5 right triangle. Notice that Sin = 0.6000.

Sin θ	*0.6000*	Csc θ	1.6666
Cos θ	0.8000	Sec θ	1.2500
Tan θ	0.7500	Cot θ	1.3333

Look at the functions table on pages 214 and 215.

To find the degrees of the reference angle, go down the sine column until you get to the number closest to 0.6000. The closest you will find is 0.6018. Reading the angle to the left of that number gives you the degrees of the reference angle, 37°.

Deg↓	Radian↓	**Sin θ↓**	Cos θ↓	Tan θ↓	Cot θ↓	Sec θ↓	Csc θ↓		
37°	0.6458	**0.6018**	0.7986	0.7536	1.3270	1.2521	1.6616	0.9250	53°
		Cos θ↑	Sin θ↑	Cot θ↑	Tan θ↑	Csc θ↑	Sec θ↑	Radian↑	Deg↑

Notice that you could have used any of the functions to find this angle.

Complementary Angle

Notice the 53° angle on the far right side of the same row. This is the other angle in the 3-4-5 triangle and is called the **complementary angle.** *A complementary angle is found by subtracting the reference angle from 90°* (90° - 37° = **53**°).

Since the reference and complementary angles are in the same triangle, both angles use the same numbers (length of sides) to derive their functions. The sine of the reference angle is equal to the cosine of the complementary angle, and vice versa.

The following charts show which functions of the reference and the complementary angles are equal to each other.

Deg↓	Radian↓	Sin θ↓	Cos θ↓	Tan θ↓	Cot θ↓	Sec θ↓	Csc θ↓		
		Cos θ↑	Sin θ↑	Cot θ↑	Tan θ↑	Csc θ↑	Sec θ↑	Radian↑	Deg↑

Notice that any two functions opposite each other are identified by a name and its **co**name.

Ref Angle Function	Comp Angle Function
Sine	**Co**sine
Cosine	Sine
Tangent	**Co**tangent
Cotangent	Tangent
Secant	**Co**secant
Cosecant	Secant

Note: If you name the sides incorrectly, you will end up with the complementary angle instead of the reference angle. One way to avoid this error is to remember that the shortest side is always opposite the smallest angle.

> To solve any math problem, it is necessary to know a certain amount of information. The information given in this book, referred to as *knowns*, is enough information to begin solving the problem presented; however, you may need to add to those knowns by further calculations to arrive at the final answer. The *needs* are the information you are trying to find.

Here is a problem which uses the table to find the angles.

Find the angles of this triangle.

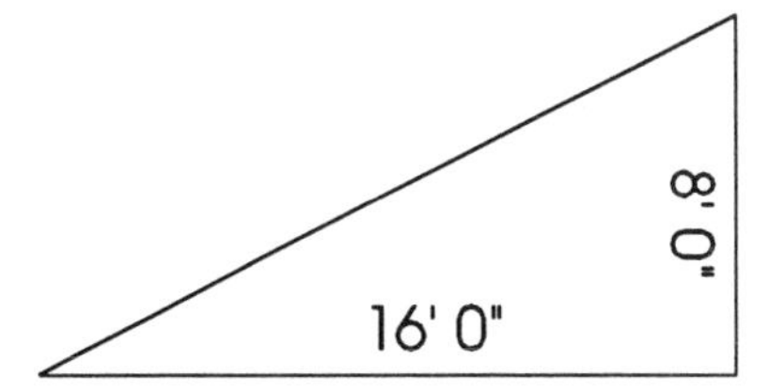

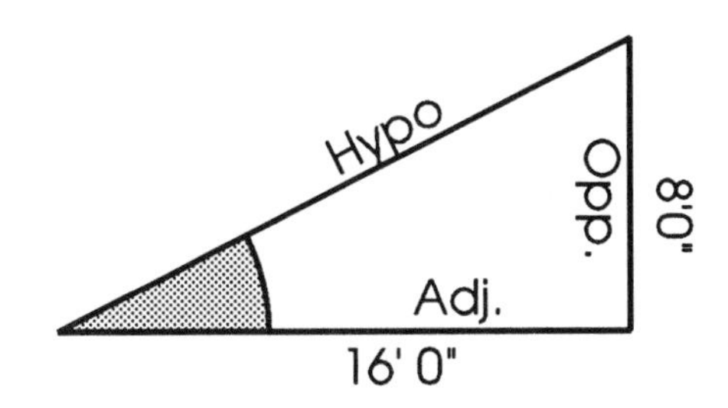

First: Draw a thumbnail sketch, then label the reference angle and name the sides.

- Adjacent side = 16' Opposite side = 8'

Second: Choose the function that uses the *knowns* to determine the *needs*. Replace the ratio with the *knowns* and calculate the problem.

Sin θ = O/H	Csc θ = H/O
Cos θ = A/H	Sec θ = H/A
Tan θ = O/A	**Cot θ** = A/O

- There are two functions which can be used when you know just the opposite and adjacent sides: tangent or cotangent.

Cotangent is used this time.

$$\text{Cotangent } \theta = \frac{\text{adjacent side}}{\text{opposite side}} = \frac{16}{8} = 2$$

Three: Refer to the function table to determine the angle.

- Look at the functions table on pages 214 and 215 under cot θ for 2.
  The closest angle is 26.5°, or **$26\frac{1}{2}$**°.
- The complementary angle is 90° - 26.5° = 63.5°, or **$63\frac{1}{2}$**°.
  Did you notice the complementary angle on the right side of the table?

Practice 26: Find the unknown angles for each right triangle below. Use the functions table on pages 214 and 215. Answers-Page 219.

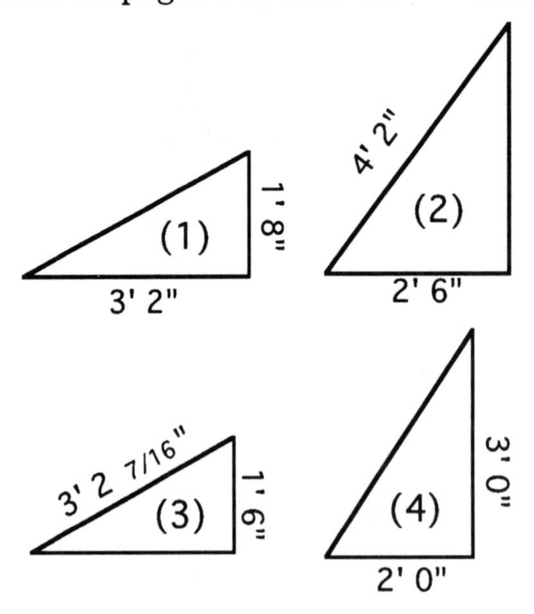

Using the Function Keys

There are three function keys on the calculator.

shows the sine of an angle in the display.

shows the cosine of an angle in the display.

shows the tangent of an angle in the display

Example:

If you need the cosine of a 45° angle, enter 45 and push the [cos] key.

Enter 45	Display [45]
Push [cos]	Display [0.707106781]

The number in the display is the cosine of a 45° angle.

Practice 27: Find the indicated function for each angle below. Round off to 4 decimal places. Answers-Page 219.

(1)	tan 45°	(5)	cos 54°
(2)	sin 33°	(6)	tan 75°
(3)	cos 25°	(7)	sin 89°
(4)	sin 60°	(8)	cos 60°

Using the Inverse Key

The fact that the calculator has keys for only half the functions and arc functions is not an oversight on the part of the manufacturers. It is pure economics. Why should they put six keys on a calculator when one key, the inverse key $\boxed{\frac{1}{x}}$, will do the same work?

Look at the symbol on the inverse key. It is a fraction with a variable as the denominator. That variable will be whatever is in the display when you push the inverse key. If you have 2 on the display and push $\boxed{\frac{1}{x}}$, the display will show 0.5, which is equal to $\frac{1}{2}$. The inverse key replaces the variable x with the number in the display and gives you the answer in the format of a decimal point fraction.

Using the Inverse Key with Functions

Let's turn to the functions and see how they work with this inverse key.

$\text{Sin } \theta = \frac{O}{H}$	$\text{Csc } \theta = \frac{H}{O}$
$\text{Cos } \theta = \frac{A}{H}$	$\text{Sec } \theta = \frac{H}{A}$
$\text{Tan } \theta = \frac{O}{A}$	$\text{Cot } \theta = \frac{A}{O}$

Notice that the numerator and denominator of the cosecant, $\frac{H}{O}$, are the reverse of the sine, $\frac{O}{H}$. They have an inverse relationship. In fact, all the functions on the right are inverse to the functions on the left, and vice versa.

For example: With the 3-4-5 right triangle, the Sin $\theta = \frac{3}{5}$ and the Csc $\theta = \frac{5}{3}$.

$$\frac{3}{5} = 0.6 \text{ and } \frac{5}{3} = 1.67$$

Enter .6 $\boxed{\frac{1}{x}}$ Display 1.666666667

Push $\boxed{\frac{1}{x}}$ Display 0.6

This chart shows these inverse relationships.

$\text{Sin } \theta = \frac{1}{\csc \theta}$	$\text{Csc } \theta = \frac{1}{\sin \theta}$
$\text{Cos } \theta = \frac{1}{\sec \theta}$	$\text{Sec } \theta = \frac{1}{\cos \theta}$
$\text{Tan } \theta = \frac{1}{\cot \theta}$	$\text{Cot } \theta = \frac{1}{\tan \theta}$

To determine a function using the inverse key:

First: *Determine the function which is inverse to the function you want.*

Second: *Do your calculations based on that function.*

Third: *Push the inverse key,* $\boxed{\frac{1}{x}}$.

For example, to find the cosecant of an angle:

First: Find the sine of the angle.

Second: Use the inverse key, $\boxed{\frac{1}{x}}$, to find the cosecant of the angle.

When you enter the sine of an angle into the calculator and push the inverse key, you have $\frac{1}{\sin\theta}$ which is the cosecant of that angle.

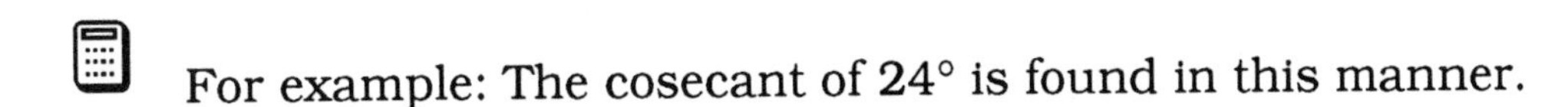

For example: The cosecant of 24° is found in this manner.

Enter 24	Display	24	
Push $\boxed{\sin}$	Display	0.406736643	(This is the sin of 24°)
Push $\boxed{\frac{1}{x}}$	Display	2.45893336	(This is the cosecant of 24°)

The cosecant of 24° is **2.4589**.

Practice 28: Find the function shown for each angle. Round each answer off to 4 decimal places. Answers-Page 219.

Remember: Use the calculator key that is the inverse of these functions first, then the $\boxed{\frac{1}{x}}$ key.

(1) csc 40°

(2) sec 45°

(3) cot 34°

(4) sec 12°

(5) cot 60°

(6) csc 24°

(7) sec 65°

(8) cot 1°

$\text{Sin}\,\theta = \frac{1}{\csc\theta}$	$\text{Csc}\,\theta = \frac{1}{\sin\theta}$
$\text{Cos}\,\theta = \frac{1}{\sec\theta}$	$\text{Sec}\,\theta = \frac{1}{\cos\theta}$
$\text{Tan}\,\theta = \frac{1}{\cot\theta}$	$\text{Cot}\,\theta = \frac{1}{\tan\theta}$

Using the Inverse Key with the Arc Functions

In the earlier section on arc functions, you used only the arc functions that were on the calculator—arc sine, arc cosine, and arc tangent. Now that you are familiar with the inverse key, you can use it to find the degrees of angles using the other three arc functions.

For example: By knowing the cotangent of an angle, you can find the degrees of the angle by following the below steps.

First: Use the inverse key, [$\frac{1}{x}$], to change the cotangent to the tangent.

Second: Push the arc tangent key, [$\tan^{-1}$]*, to find the degrees.

Example: Find the angle that has a cotangent of 0.7002.

Enter .7002	Display	0.7002	(Cotangent)
Push [$\frac{1}{x}$]	Display	1.428163382	(Tangent)
Push [$\tan^{-1}$]	Display	55.00028982	(Degrees in angle)

The answer when rounded off to the nearest $\frac{1}{2}$° is **55°**.

Practice 29: Find the angle for each function. Round the answers off to the nearest $\frac{1}{2}$ degree. Answers-Page 219.

Remember: The [$\frac{1}{x}$] key is used with functions and not degrees. Do not be a speed demon on the calculator. Wait until the function shows up on the display before you push the [$\frac{1}{x}$] key.

(1)	$\cot\theta$	=	0.1777	(5)	$\sec\theta$	=	2.2034
(2)	$\csc\theta$	=	1.0302	(6)	$\cot\theta$	=	0.9014
(3)	$\sec\theta$	=	1.1040	(7)	$\csc\theta$	=	2.622
(4)	$\cot\theta$	=	12.154	(8)	$\sec\theta$	=	5.1216

* Remember, on your calculator you may need to use the [shift] [tan] or [2nd] [tan] keys instead of the [$\tan^{-1}$] key. Refer to your calculator manual.

Calculating Right Triangles Using One Side and One Angle

So far we have used the functions only to determine angles of the right triangle. We are now going to use those same functions to find the unknown sides of the right triangle when one side and one angle (besides the right angle) are known.

A method for determining leg lengths can be found by rearranging the functions and the ratios. *You can multiply both sides of a formula by the same number or function and the formula will remain equal.*

For example, $\text{Cosecant } \theta = \dfrac{\text{Hypotenuse}}{\text{Opposite side}}$

If you multiply both sides of the cosecant ratio by the opposite side (denominator of the ratio), you can determine a formula to find the length of another side.

$$\text{Opposite} \times \text{Cosecant } \theta = \frac{\text{Hypotenuse}}{\cancel{\text{Opposite}}} \times \cancel{\text{Opposite}}$$

The two *opposites* on the right side of the equal mark cancel each other out and leave us with this formula:

$$\text{Opposite side} \times \text{Cosecant } \theta = \text{Hypotenuse}$$

This formula indicates that the length of the opposite side (of a right triangle) times the cosecant (of the reference angle) equals the length of the hypotenuse.

All of the functions can be rearranged using this method.

$$\text{Hypotenuse} \times \text{Sine } \theta = \text{Opposite side}$$
$$\text{Opposite side} \times \text{Cosecant } \theta = \text{Hypotenuse}$$
$$\text{Hypotenuse} \times \text{Cosine } \theta = \text{Adjacent side}$$
$$\text{Adjacent side} \times \text{Secant } \theta = \text{Hypotenuse}$$
$$\text{Adjacent side} \times \text{Tangent } \theta = \text{Opposite side}$$
$$\text{Opposite side} \times \text{Cotangent } \theta = \text{Adjacent side}$$

To make the chart easier to read, the terms on the chart on the next page have been placed in a different order. This chart will be referred to as the *NAK chart.*

NAK Chart				
Need		**Angle**		**Known**
Hyp	=	csc θ	x	Opp
Hyp	=	sec θ	x	Adj
Opp	=	tan θ	x	Adj
Opp	=	sin θ	x	Hyp
Adj	=	cos θ	x	Hyp
Adj	=	cot θ	x	Opp

The NAK chart shows which functions (of the angle) should be multiplied by the *known* side to get the needed side.

Notice how each side of a right triangle can be found if either of the other sides and one angle (other than the 90°) are known.

Remember: Use the NAK chart if you know one angle (other than the right angle) and the length of one side.

Example: A right triangle has a hypotenuse with a length of 4' and an angle of 35°, and you want to determine the length of the other sides.

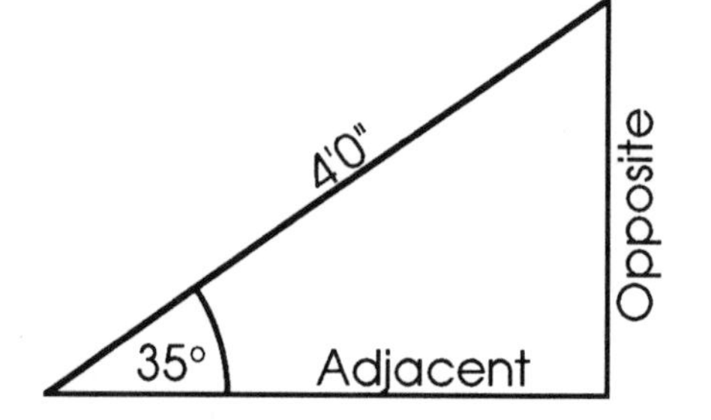

First: Determine your *knowns* and *needs*.

The *knowns* are the hypotenuse of 4" and the reference angle of 35°. The *needs* are the opposite and adjacent sides.

Second: Use the NAK chart to determine which functions use the known side to find the needed sides.

Notice that there is one function for each needed side that uses the known side.

NAK Chart				
Need		**Angle**		**Known**
Hyp	=	csc θ	x	Opp
Hyp	=	sec θ	x	Adj
Opp	=	tan θ	x	Adj
Opp	=	sin θ	x	Hyp
Adj	=	cos θ	x	Hyp
Adj	=	cot θ	x	Opp

Let's first find the opposite side:

Opposite side = sin θ x hypotenuse.

Opposite side = sin 35° x 4'

Sin 35° = 0.5736

Opposite side = 0.5736 x 4'

Opposite side = **2.2943'**

Using the calculator

Enter 35 [sin]	Display	.573576436
[x] 4 [=]	Display	2.294305746

The length of the side opposite the reference angle is 2.2943', or **2' $3\frac{9}{16}$"**.

To find the length of the adjacent side when the length of the hypotenuse and an angle are known, use the cosine of the angle.

Adjacent side = Cos θ x Hypotenuse

Adjacent side = Cos 35° x 4'　　　　Cos 35° = 0.8192

Adjacent side = 0.8192 x 4'

Adjacent side = 3.277

Enter 35 [cos]　　　　Display 0.819152044

[x] 4 [=]　　　　Display 3.276608177

The length of the adjacent side is 3.277', or **3' $3\frac{5}{16}$"**.

Practice 30: Find the lengths of the two unknown sides (the angle given is the reference angle for the side). Round off your answers to 4 decimal places. Answers-Page 219.

	Deg	Side	Lg		Deg	Side	Lg
(1)	65°	opp	4"	(5)	37°	adj	6'
(2)	45°	hypo	5'	(6)	55°	opp	3'
(3)	20°	adj	12"	(7)	70°	hyp	136"
(4)	1°	opp	1"	(8)	88°	hyp	99'

NAK Chart

Need		Angle		Known
Hyp	=	csc θ	x	Opp
Hyp	=	sec θ	x	Adj
Opp	=	tan θ	x	Adj
Opp	=	sin θ	x	Hyp
Adj	=	cos θ	x	Hyp
Adj	=	cot θ	x	Opp

The functions may be found with the calculator or the tables on pages 214-215.

Note: You now have two ways of finding the sides of a right triangle.

1. If you know two sides, you can use the Pythagorean Theorem.
2. If you know one angle and the length of one side, you can rearrange the functions and ratios or use the NAK chart.

notes

IX

Isosceles Triangles

An isosceles triangle is *a triangle with two equal sides and two equal angles.* The easiest method to work an isosceles triangle is to divide it into two equal right triangles. Since you know how to work right triangles, you also have a key to working isosceles triangles.

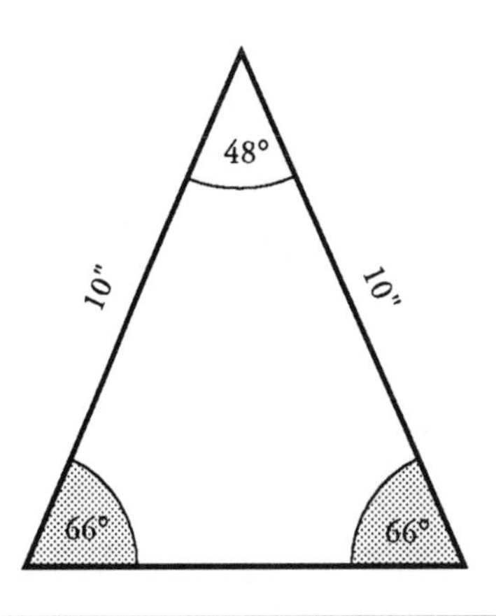

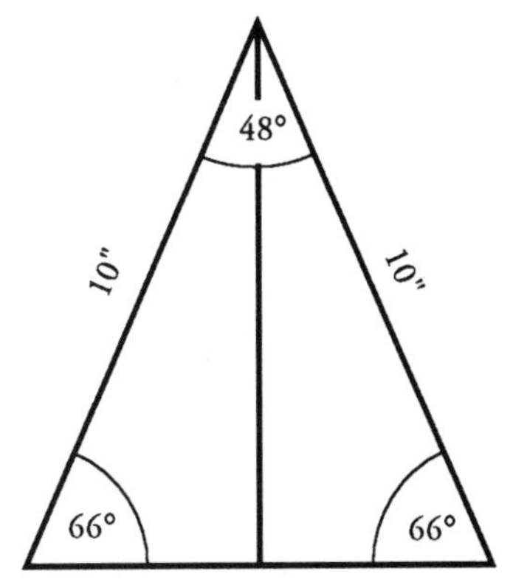

*A line that divides an angle, line, or triangle into equal parts is said to **bisect** them.*

The bisecting line of an isosceles triangle drops from the vertex of the odd angle to the base. It divides the odd angle, the base*, and the isosceles triangle into two equal parts. The bisecting line of an isosceles triangle is also perpendicular to the base. Therefore, the two triangles created are right triangles.

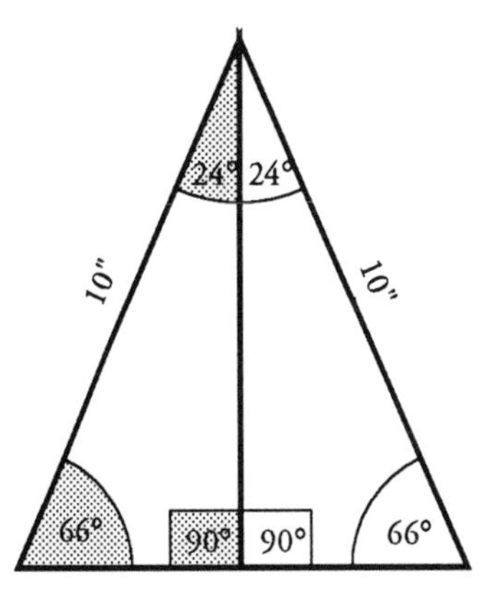

Solve the missing angles and side lengths for one right triangle and you have the information needed to find the sides of the isosceles triangle.

* The base of a triangle is the side that it sits on. The base of an isosceles triangle is usually the odd side.

Example: Use the 24° angle (of the bisected isosceles triangle) as the reference angle. Find the unknowns parts of the isosceles triangle.

First: Define your *knowns* and your *needs*.

- The ***knowns*** are: A 24° reference angle and a 10" hypotenuse.

The ***need*** is: The length of the opposite side

Remember: *Think NAK chart if you know one side and one angle.*

Second: Look at the NAK chart and decide which function allows you to find the *needs* by using the *knowns*.

$$\text{Opp} = \text{Sin}\ \theta \times \text{Hyp}$$

Third: Put in the *knowns* and calculate the problem. $\theta = 24°$ Hyp = 10"

NAK Chart

Need		Angle		Known
Hyp	=	csc θ	x	Opp
Hyp	=	sec θ	x	Adj
Opp	=	tan θ	x	Adj
Opp	=	sin θ	x	Hyp
Adj	=	cos θ	x	Hyp
Adj	=	cot θ	x	Opp

Opposite side = Sin θ x Hypotenuse

Opposite side = sin 24° x 10"

Opposite side = 0.4067 x 10"

Opposite side = 4.067"

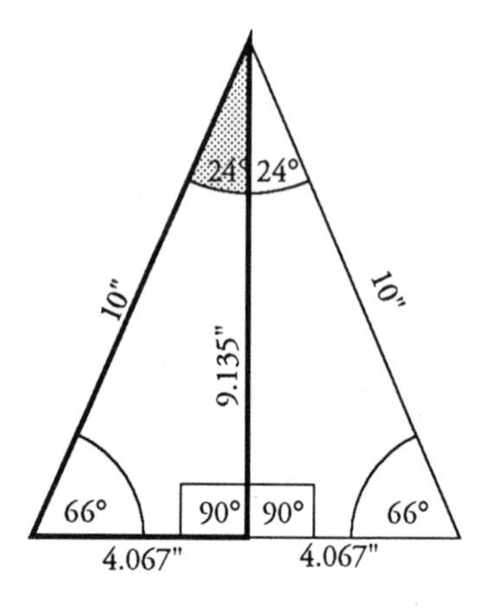

Since the right triangle is $\frac{1}{2}$ of the isosceles triangle, the length of the base of the isosceles triangle is twice the length of the base of the right triangle. 2 x 4.067" = 8.134"

Seeing is Believing
Page 194

Remember: To find the two equal angles in the isosceles triangle, subtract the odd angle from 180°, then divide the answer by two. To find the odd angle, subtract the sum of the two known angles from 180°.

Practice 31: Find the length of the base and the equal angles for each isosceles triangle below. Round the answers off to four decimal places. Answers-Page 219.

	Odd angle	Equal sides		Odd angle	Equal sides
(1)	60°	22"	(5)	90°	14'
(2)	23.5°	17'	(6)	75°	4.5"
(3)	45°	62.245"	(7)	18°	41"
(4)	72°	54"	(8)	30°	78'

notes

Special Triangles

The **45° isosceles right triangle** and **30°- 60° right triangle** are used in the field every day. Their functions are used as *constants* by different crafts. The constant 1.4142 is used by pipe fitters when they work a 45° offset. Electricians use a constant of 2 when they work a 30° offset. Both constants are functions of the angle of the offset they are using. Let's look at these right triangles in more detail.

The 45° Right Triangle

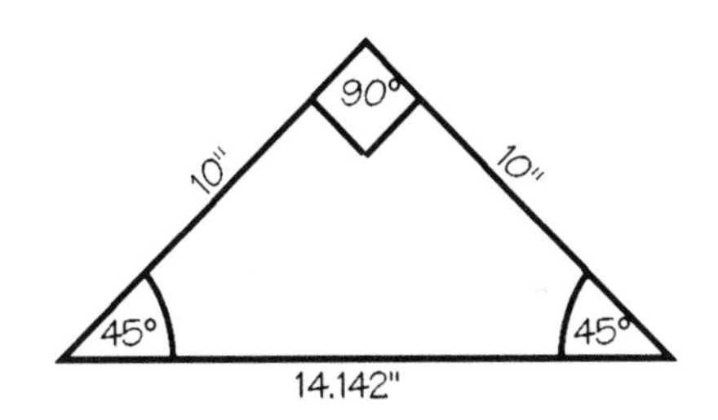

Remember: A diagonal line across a square creates two equal 45° right triangles.

Since the 45° right triangle has two equal angles and two equal sides, it is an isosceles triangle. In fact, *the 45° right triangle is the only right triangle that is an isosceles triangle.*

An interesting thing happens to this isosceles triangle when a bisecting line is dropped from the vertex of the odd angle to the midpoint of the base line. Two more equal 45° isosceles right triangles are formed. Notice that the length of the bisecting line is the same length as half of the base, 7.071".

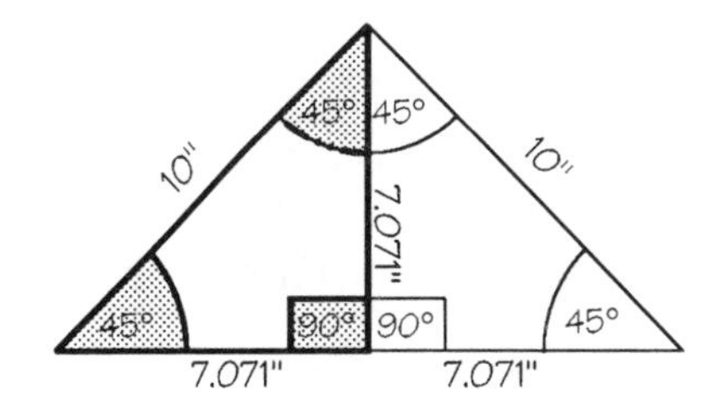

Remember: ***The legs of a 45° right triangle are and must be equal.*** You cannot have a 45° right triangle with unequal legs. Many mistakes have been made in the field because some did not understand this.

To learn more about the 45° right triangle, let's look at a 45° right triangle with equal legs of one inch in length. We are going to find the length of the hypotenuse using the Pythagorean Theorem, then determine the ratio of the sides.

$$a^2 + b^2 = c^2$$
$$\sqrt{a^2 + b^2} = c$$
$$\sqrt{1^2 + 1^2} = c$$
$$\sqrt{1 + 1} = c$$
$$\sqrt{2} = c$$
$$1.4142'' = c$$

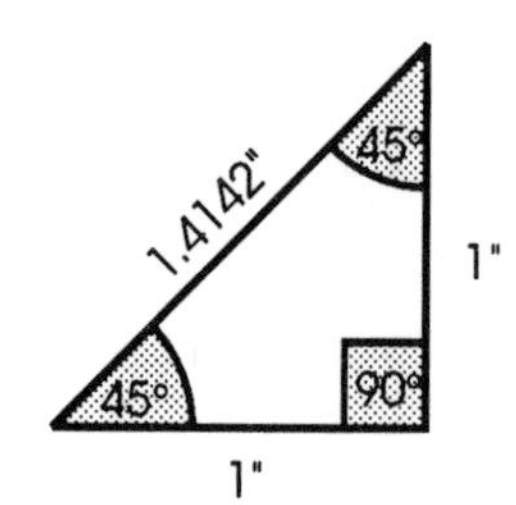

A quick way to enter 1.4142 in your calculator is to enter 2, then press [√]

$$\text{Sin } 45° = \frac{\text{opp}}{\text{hypo}} = \frac{1}{1.4142} = .7071$$
$$\text{Cos } 45° = \frac{\text{adj}}{\text{hypo}} = \frac{1}{1.4142} = .7071$$
$$\text{Tan } 45° = \frac{\text{opp}}{\text{adj}} = \frac{1}{1} = 1$$
$$\text{Csc } 45° = \frac{\text{hypo}}{\text{opp}} = \frac{1.4142}{1} = 1.4142$$
$$\text{Sec } 45° = \frac{\text{hypo}}{\text{adj}} = \frac{1.4142}{1} = 1.4142$$
$$\text{Cot } 45° = \frac{\text{adj}}{\text{opp}} = \frac{1}{1} = 1$$

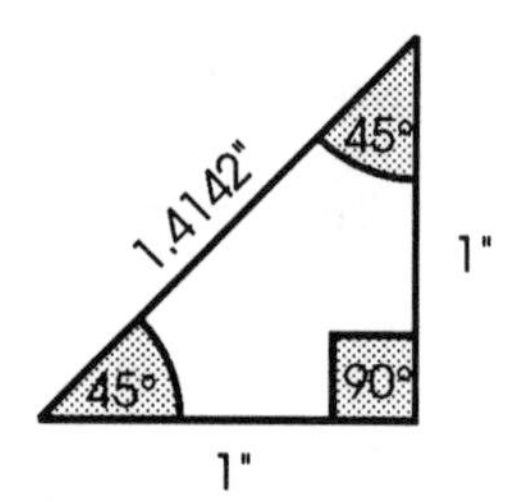

Here are the functions for a 45° angle in the NAK chart.

Need		45° Angle		Known
Hypotenuse	=	**Csc 45° = 1.4142**	**x**	**Opposite side**
Hypotenuse	=	**Sec 45° = 1.4142**	**x**	**Adjacent side**
Opposite side	=	Tan 45 = 1	x	Adjacent side
Opposite side	=	**Sin 45° = 0.7071**	**x**	**Hypotenuse**
Adjacent side	=	**Cos 45° = 0.7071**	**x**	**Hypotenuse**
Adjacent side	=	Cot 45° = 1	x	Opposite side

Notice that the length of the hypotenuse can be found by multiplying either the adjacent or opposite sides by **1.4142.**

Notice that either leg can be found by multiplying the length of the hypotenuse by **0.7071.**

This is why pipe fitters use the constants 1.4142 and .7071 with 45° offsets.

Practice 32: Find the hypotenuse or the leg length of a 45° right triangle when the other is known. Round your final answers off to four decimal places. Answers-Page 219.

	Leg	Hypo		Leg	Hypo
(1)	7"		(5)	63.125"	
(2)		17'	(6)		43'
(3)	23"		(7)	75.5"	
(4)		31.5625"	(8)		88.875"

Seeing is Believing
Page 188

The Equilateral Triangle

The **equilateral* triangle** *has three equal angles and three equal sides.* If you bisect the equilateral triangle, you create two 30°- 60° right triangles.

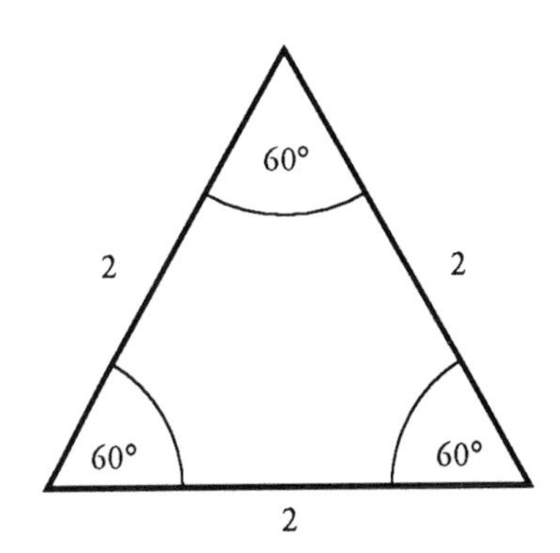

The 30°- 60° Right Triangle

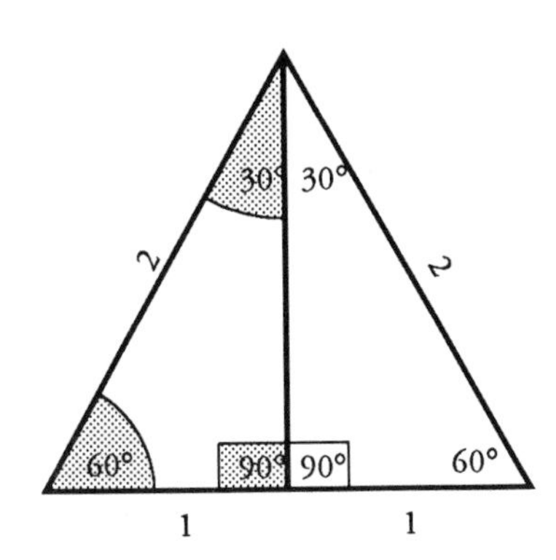

A bisecting line dropped from the vertex of any angle of an equilateral triangle to the midpoint of the opposite side creates two **30°- 60° right triangles.**

Craftsmen have known for years about the 30°- 60° right triangle. This triangle's longest side is twice as long as its shortest side. Of course, the shortest side is half as long as the longest side. This is the simplest comparison of sides we have. It is quite simple to determine that the shortest side of a 30°- 60° right triangle is 6' if you know the longest side is 12'.

Look at the ratio of sides for a 30°- 60° triangle as shown in the function table. Notice the 30° angle and the complementary angle.

Deg↓	Radian↓	Sin θ ↓	Cos θ ↓	Tan θ ↓	Cot θ ↓	Sec θ ↓	Csc θ ↓		
30°	0.5236	0.5000	0.8660	0.5774	1.7321	1.1547	2.0000	1.0472	60°
		Cos θ ↑	Sin θ ↑	Cot θ ↑	Tan θ ↑	Csc θ ↑	Sec θ ↑	Radian ↑	Deg↑

* Equilateral refers to the triangle having equal sides. The triangle can also be called an equiangular triangle since it has equal angles.

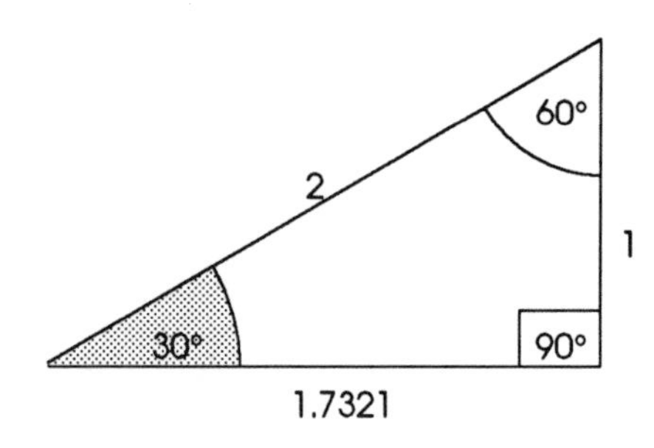

When the length of the hypotenuse is divided into the shortest side (Sin), the ratio is $\frac{1}{2}$, or .5. When the shortest side is divided into the length of the hypotenuse (Csc), the ratio is $\frac{2}{1}$, or 2.

Look at the calculations using the NAK chart.

Need		30° Angle		Known
Hypotenuse	**=**	**Csc 30° = 2**	**x**	**Opposite side**
Hypotenuse	=	Sec 30° = 1.1547	x	Adjacent side
Opposite side	=	Tan 30° = 0.5774	x	Adjacent side
Opposite side	**=**	**Sin 30° = 0.5**	**x**	**Hypotenuse**
Adjacent side	=	Cos 30° = 0.8660	x	Hypotenuse
Adjacent side	=	Cot 30° = 1.732	x	Opposite side

←Notice that the length of the opposite side times two equals the length of the hypotenuse.

←Notice that the length of the hypotenuse times .5 equals the length of the opposite side.

Practice 33: Find the length of the unknown side for these 30°- 60° right triangles. Round the final answers off to four decimal places. Answers-Page 219.

	Opp	Hypo		Opp	Hypo
(1)	34'		(4)		16'
(2)		17"	(5)	56.25"	
(3)	46"		(6)		78.5625"

Something you will see later:

The 30°- 60° right triangle is often used in crafts that require the bending of conduit, pipe, and tubing. The shortest side of a 30°- 60° right triangle is exactly half the length of the longest side (hypotenuse). This means that if a line needs to rise 11" over an obstruction, the run will be 22" in length. This drawing shows the conduit being brought back to the original elevation after the obstacle has been passed. This procedure is called a return offset.

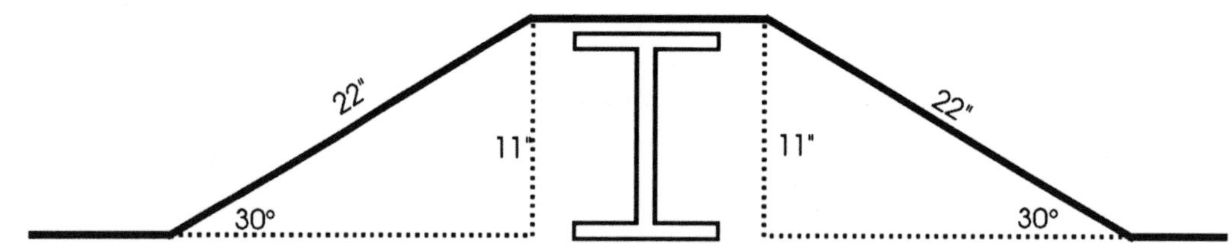

XI

Circles

A **circle** is a closed curved line around a **center point** on which every point on the curved line is exactly the same distance from the center.

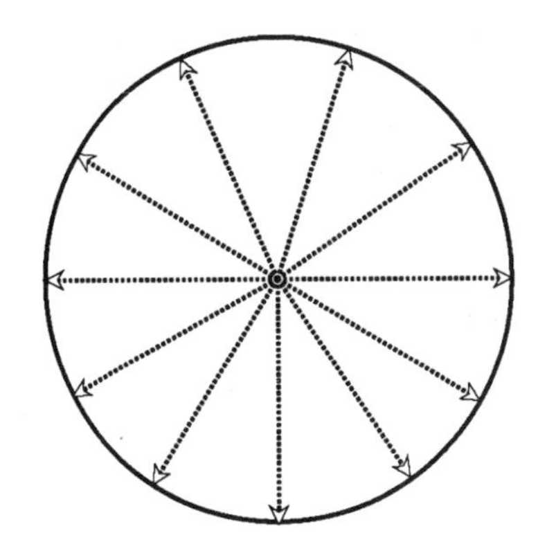

Here are some definitions that you will be using in this section on circles. Read through them now and return to them as you need them.

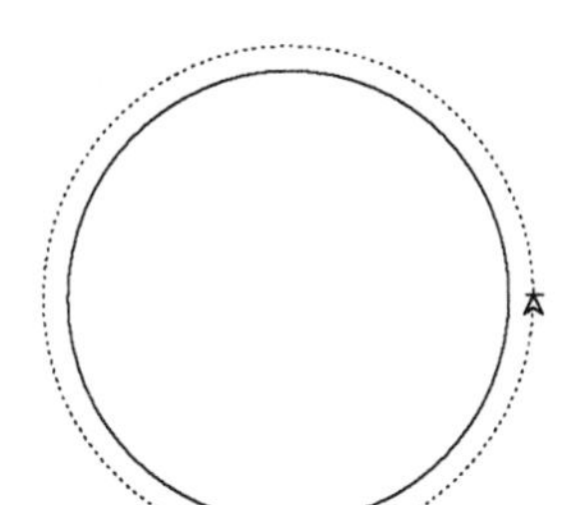

The **circumference** is the *length* of the closed curved line that forms the circle. *Circumference is abbreviated as* **c**.

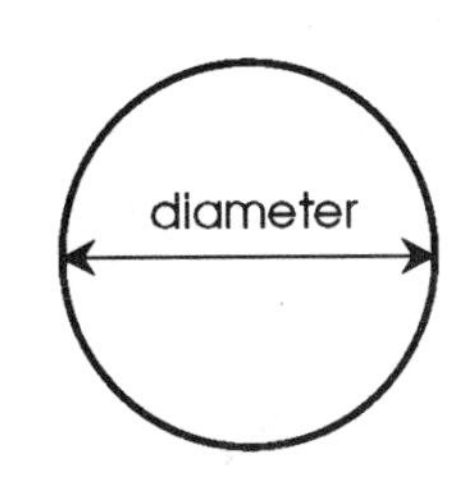

The **diameter** is the length of the straight line that goes from one point of a circle to another point of the circle and passes through the center. It is the longest straight line within a circle.

Diameter is abbreviated as ***d***.

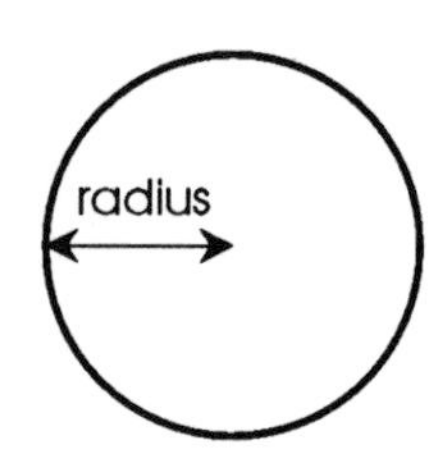

The **radius** is the length of the straight line that goes from the center point of the circle to any point on the circle.

$\frac{\text{Diameter}}{2}$ = radius. Radius x 2 = diameter.

The plural of radius is radii. One *radius*, but two or more *radii*.
Radius is abbreviated as r.

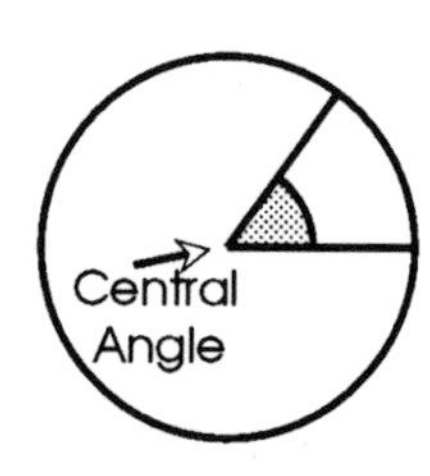

A **central angle** is an angle with the vertex located at the center point of a circle.
The central angle of a whole circle is 360°. A central angle of less than 360° is the central angle of an arc.

An **arc** is an unclosed curved line or a fraction of a circle. Like the circle, it has a radius and a central angle. The arc is referred to by its angle. This is a 70° arc.

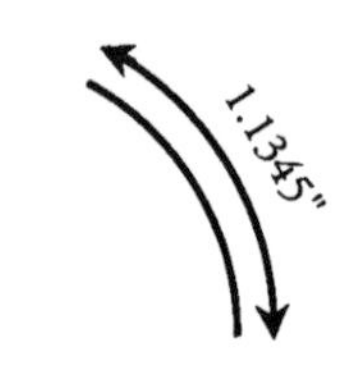

An **arc length** is the length of the arc. It is a fraction of the circumference of the circle.

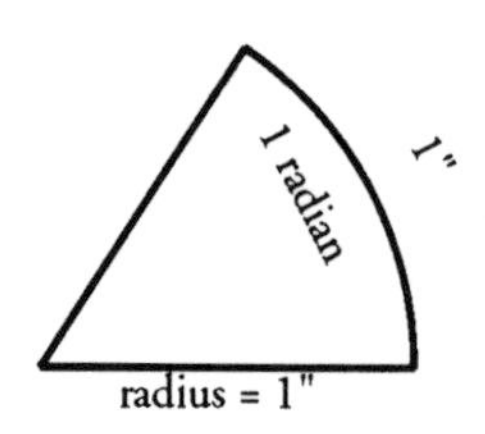

A **radian** is a measurement of arc length based on the length of the radius. One radian has an arc length equal to the length of the radius. If you divide the radius into the circumference, you find that there are 6.2832 radians in a circle. That is equal to 2 x π. One degree is equal to 0.01745 radians.

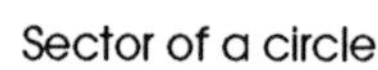
Sector of a circle

A **sector** is a portion or a fraction of a circle. It includes the arc and the two radii that connects the ends of the arc to the center of the circle. The two radii create a central angle since the vertex is in the center of the circle. Sectors are referred to by their angle. The drawing shows a 60° sector of a circle.

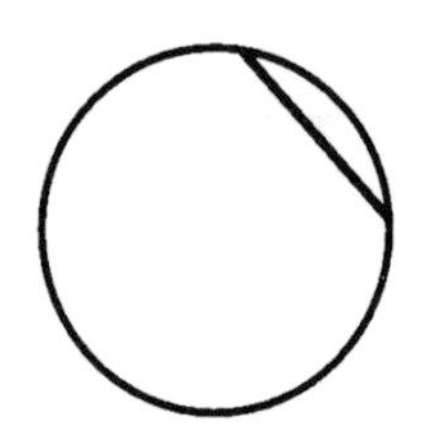

A **chord** is a straight line from one point on a circle to another point on the circle. The diameter is the longest chord of a circle.

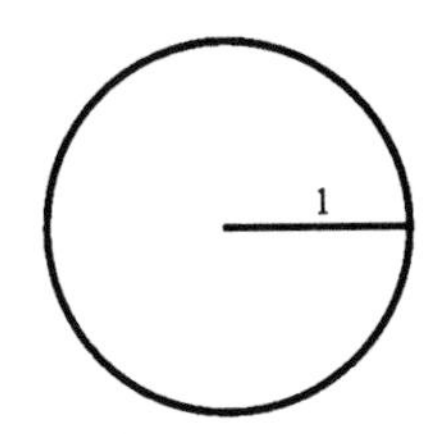

A **unit circle** is a circle with a radius of 1 unit. It is helpful in explaining not only the circle and its parts, but also the right triangle.

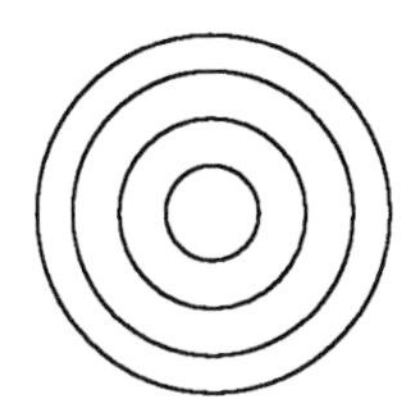

Concentric circles have the same center point, but different radii lengths.

Concentric arcs have the same center point, but different radii lengths. Most of the concentric arcs you see will also have the same arc angle.

PI π

The symbol used when dealing with round objects is **pi(π).** π is a letter in the Greek alphabet which is used in mathematics to represent a particular number. To a mathematician, it is the whole number 3 and an endless number of decimal places. Scientists working on space travel may use a whole page of decimal places after the number 3. The calculator key [π] will display 3.141592654, but most of us remember π from our school days as 3.14.

π is the number that when multiplied by the diameter of a circle gives you the circumference of that circle. If you take a 1" circle and unroll it, it will measure approximately 3.1416".

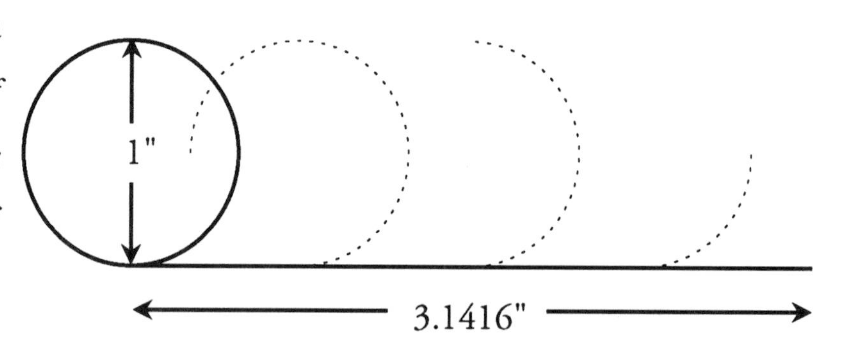

In fact, early mathematicians rolled out all sizes of circles, and when they divided the diameter of the circle into the length rolled out, the same number, 3.14, always appeared. The number was called pi.

Don't be confused by all the decimal places. Use whatever is in your calculator, and round the answer off to serve your need for accuracy. If you are using a calculator that doesn't have a π key, use 3.1416.

Circumference

Using the Diameter to Find the Circumference

The distance around a circle is the **circumference** *of the circle.* From the section on pi, you know that the circumference of a circle is found by multiplying the diameter of the circle by π.

The formula is: **$c = \pi d$**

Here is an example of calculating circumference using a calculator.

Find the circumference of a circle that has a diameter of 6.

Push [π] [x]	Display	3.141592654
Enter 6 [=]	Display	18.84955592

The circumference of a circle with a diameter of 6 is 18.8496. The unit of measurement could be inches, foot, meter, mile, or whatever unit you are using. If you are using inches, then the circumference is $18\frac{7}{8}$", but if you are using feet, the circumference is 18'$10\frac{3}{16}$".

Practice 34: Find the circumference of these circles. Convert the decimals to fractions. Answers-Page 219.

(1) d = 3'

(2) d = 4.5"

(3) d = $10\frac{7}{8}$"

(4) d = 7.75'

(5) d = 9"

(6) d = $3\frac{3}{4}$"

(7) d = 16'

(8) d = 28'

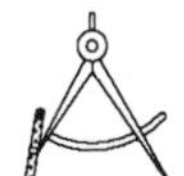

Seeing is Believing: Page 195

Using the Radius to Find the Circumference

The **radius** *is* $\frac{1}{2}$ *the length of the diameter.* $\boxed{\mathbf{r = \frac{d}{2}}}$ **or** $\boxed{\mathbf{d = 2r}}$. If the diameter is 4, then the radius is 2. If the radius is 3, then the diameter is 6.

To find the circumference of a circle using the radius instead of the diameter, replace the formula $\mathbf{c = \pi d}$ with the formula $\mathbf{c = \pi 2r}$.

This formula is accurate since $\mathbf{d = 2r}$. (You will most often see this formula written as $\mathbf{2\pi r}$; however, the order of the numbers or variables does not matter when multiplying.)

Example: Find the circumference of a circle that has a radius of 5".

Push [π] [x]	Display	3.141592654
2 [x]	Display	6.283185307
5 [=]	Display	31.41592654

The circumference of a circle with a 5" radius is $31\frac{7}{16}$".

Practice 35: Find the circumference for each circle. Convert the answers to the appropriate whole units and fractions. Answers-Page 219.

(1) r = 2"

(2) r = 15'

(3) r = 19"

(4) r = 123'

(5) $r = 3\frac{1}{8}$ "

(6) r = 2'

(7) $r = 15\frac{1}{4}$ "

(8) r = 50'

Arc

You might call an *arc* an incomplete or unfinished circle. If you use a compass and start drawing a circle but don't finish, you will have drawn a part, or a fraction, of a circle.

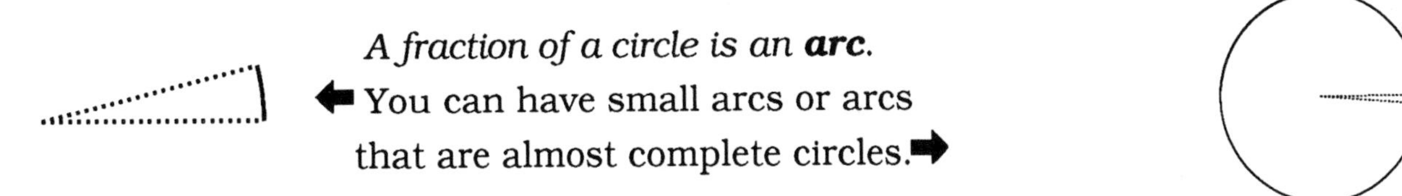

> Every point on an arc is the same distance from the center point (just as with a circle). The **radius of the arc** is the same as the **radius of the circle** it is a part of.

The Arc as a Fraction of a Circle

Since an arc is a part of a whole circle, one way to calculate the measurements of an arc is to know what fraction of the whole circle the arc is, and multiply that fraction by the circumference to find arc length or 360° to find the angle of the arc.

For instance: If an arc is equal to $\frac{1}{2}$ of the total circle, the degrees of the angle of the arc are equal to $\frac{1}{2}$ of the total degrees of the whole circle (360°).

$$\frac{1}{2} \times 360° = 180°$$

In order to set up a fraction, you must be able to compare a part of a unit to a whole unit. There are two comparisons that can be made to find what fraction of a circle the arc is.

- The first is to compare the degrees of the arc angle (the part) to the degrees of the circle (the whole).
- The second is to compare the arc length (the part) to the circumference (the whole).

Let's first look at the *degrees of the central angle of the arc compared to the total degrees of the circle* (360°).

For example: An arc has an angle measurement of 14°.

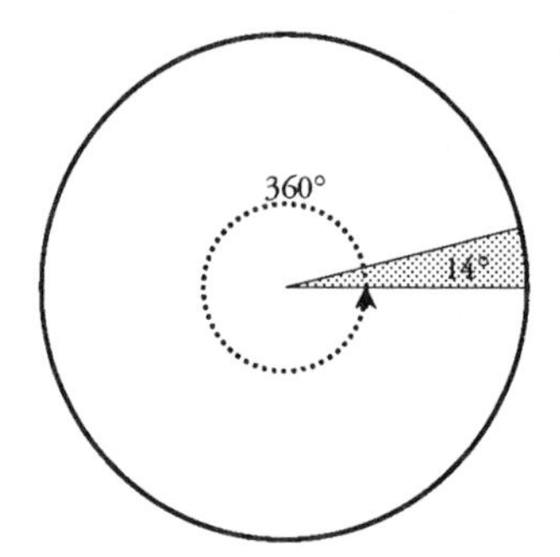

The whole circle has 360 parts and this arc has 14 of those parts.

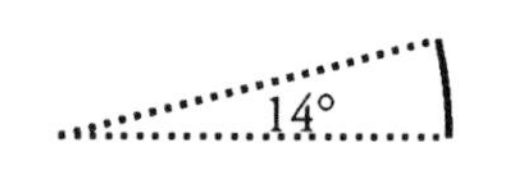

Remember: The denominator of a fraction is the number of parts a whole has been divided into, and the numerator is the number of the parts you have.

This arc is equal to $\frac{14}{360}$ of the whole circle.

Now let's look at the *length of the arc compared to the circumference of the circle.*

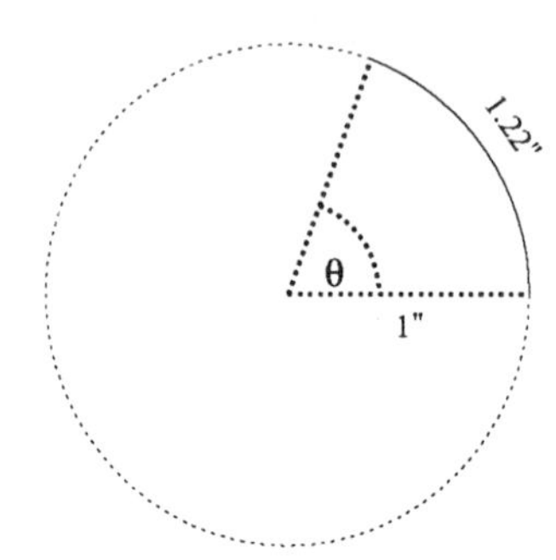

For example: Find what fraction of the circle this arc is. The arc length is 1.22" and the radius is 1".

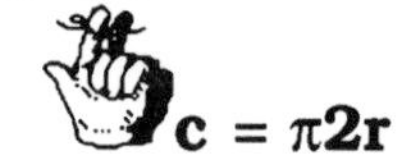

$$\text{Fraction of circle} = \frac{\text{arc length}}{\text{circumference}}$$

$$\text{Fraction of circle} = \frac{1.22"}{2\pi r}$$

$$\text{Fraction of circle} = \frac{1.22}{6.2832}$$

Practice 36: Use the stated *knowns* to determine what fraction of the circle each arc is. Answers-Page 220.

Remember: The fraction of the circle that the arc is equal to can be found by these two methods:

$$\frac{\text{Arc Length}}{\text{Circumference}} \text{ or } \frac{\text{Degrees in Arc Angle}}{360°}$$

	Arc Length	Circumference	Radius	Degrees in Arc Angle
1.	4"	12"		
2.				90°
3.	12"		2'	
4.				12°

It is important to understand that no matter which comparison is used to find the fraction of the arc, once that fraction is known, it applies to all aspects of the arc.

For instance, you know the first arc we worked with is $\frac{14}{360}$ of the whole circle. For that arc:

Arc Length = $\frac{14}{360}$ times the circumference of the circle.

Arc Angle = $\frac{14}{360}$ times the total degrees of the circle (360°).

Finding Arc Length Using a Fraction of a Circle

Let's take a closer look at finding arc length when you know the fraction of the arc is $\frac{14}{360}$.

You know that arc length = $\frac{14}{360}$ x circumference of the circle. To find the length of the arc:

First: Find the circumference of the circle.

Let's make the radius of this arc 1.125".

$$c = 2\pi r$$
$$c = 2 \times 3.1416 \times 1.125''$$
$$c = 7.0686''$$

Second: Multiply the fraction by the circumference.

Arc length = Fraction of circle x circumference of circle

Arc length = $\frac{14}{360} \times 7.0686''$ = $\frac{\text{arc angle}}{360} \times 2\pi r$ (circumference)

Arc length = $\frac{14}{360} \times \frac{7.0686}{1}$ = $\frac{\text{arc angle}}{360} \times \frac{2\pi r}{1}$

Arc length = $\frac{98.9604''}{360}$ = $\frac{\text{arc angle} \times 2\pi r}{360}$

Arc length = 0.27489"

The length of a 14° arc with a radius of 1.125" is 0.275".

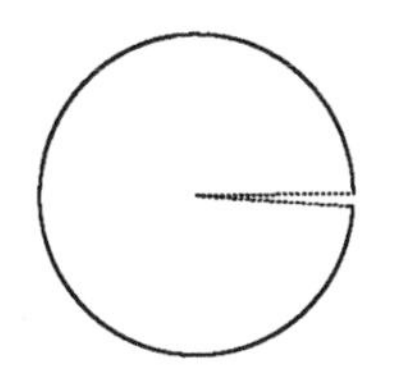

Example: Use this formula to find arc length.

The arc angle is 355° and the radius is 0.5".

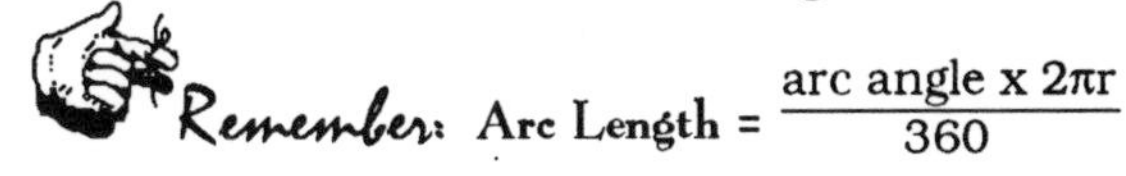

Enter	355 [x] 2 [x]	Display	710
Push	[π] [x] .5 [=]	Display	1115.268
Push	[÷] 360 [=]	Display	3.0979666

The arc on the drawing above is 3.098" long.

Practice 37: Calculate the length for each arc using the fraction of the circle. Answers-Page 220.

	° of angle	radius		° of angle	radius
(1)	45°	10"	(5)	90°	2.5'
(2)	$22\frac{1}{2}°$	12"	(6)	135°	0.5
(3)	15°	10'	(7)	85°	2.5'
(4)	10°	15"	(8)	$154\frac{1}{2}°$	$725\frac{1}{2}$"

Finding the Arc Angle Using a Fraction of a Circle

To find the degrees of an arc, multiply the fraction of the circle by the degrees in a whole circle, 360°.

For an example, let's use an arc length of 1.22" and a circumference measurement of 6.2832.

Arc angle = fraction of circle x 360°

Arc angle $= \frac{\text{arc length}}{\text{circumference}} \times 360° \quad = \frac{1.22}{6.2832} \times 360°$

Arc angle $= \frac{\text{arc length}}{\text{circumference}} \times \frac{360°}{1} \quad = \frac{1.22}{6.2832} \times \frac{360}{1}$

Arc angle $= \frac{\text{arc length} \times 360°}{\text{circumference}} \quad = \frac{1.22 \times 360}{6.2832}$

Arc angle = *69.9°*

Find the angle of an arc with a 7.25" length and a 11" radius.

Enter 2 [x] [π] [x] 11 Display 69.1152 = Circumference

Put 69.1152 in memory or write it down

Enter 7.25 [x] 360 [=] Display 2610

Push [÷] 69.1152 [=] Display 37.763039

Practice 38: Find the angle of each arc using the fraction of the circle. Round the answer to the nearest $\frac{1}{2}°$. Answers-Page 220.

	radius	arc length		radius	arc length
(1)	3"	6.28"	(5)	15"	3.9'
(2)	2'	12'	(6)	110'	12.5'
(3)	1.5'	1.36'	(7)	5.45"	3.45"
(4)	0.5"	0.26"	(8)	2.5 miles	0.7 miles

Concentric Arcs with the Same Angle

We use **concentric arcs** of the same degree more often than single arcs. The reason is simple. All of the material we work with has width. When a turn is made with an arc, there are at least three arcs involved—**inside arc**, **center arc**, and **outside arc**.

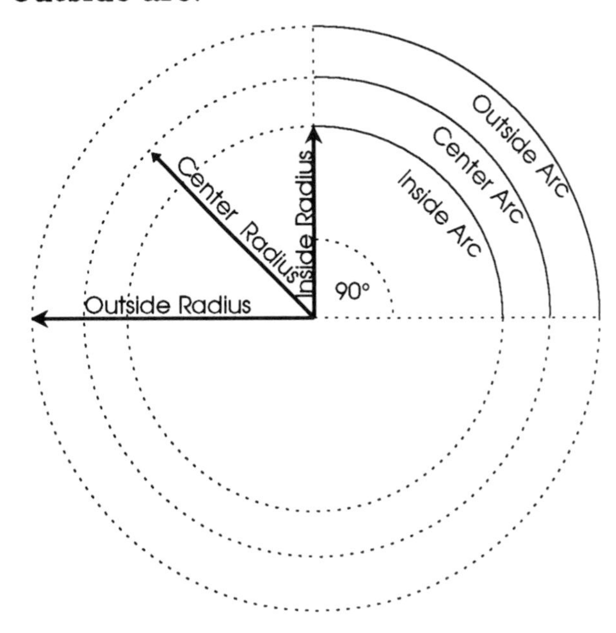

There are usually two pieces of information given about arcs on blueprints—the length of the **center radius and the degree of the arc**. These facts are enough. The rest can be calculated if you know the width of the material used in the arc. *The inside and outside radii are based on the length of the center radius plus or minus half the width or diameter of the material being used.*

In the last section, you learned one method of finding an arc length when the angle and the radius are known. With concentric arcs with the same arc angle but different radii, there is an easier method.

Unit Circle

We often calculate problems using the "*based on **1***" concept. If I were to tell you that my car will travel 30 miles on **1 gallon** of gas, then ask you how far I could travel on 5 gallons, you would multiply 5 by 30 and reply 150 miles. If I were to tell you that I can walk 3 miles in **1 hour,** then ask you how far I could walk in 3 hours, you could easily reply 9 miles.

In each case, the measurement given is "*based on **1**.*" Once you know what the measurement of the "1" is (the examples above are **1** gallon of gas and **1** hour), you can easily calculate distances based on other variables. The same type of logic can be used with the circle.

Since an arc is part of the circumference of a circle, let's look at the formula for circumference of a circle using the radius: $C = 2\pi r$.

Remember: The term for a circle with a *radius of one* is **unit circle.**

Since the radius of a unit circle is **1**, replace the **r** in the circumference formula with 1.

$$C = 2\pi r \text{ or } \pi \times 2 \times r$$
$$C = \pi \times 2 \times 1$$
$$C = \pi \times 2 \text{ or } 2\pi$$
$$C = 2\pi \text{ or } 6.2832$$

The circumference of the unit circle is 2π or 6.2832.

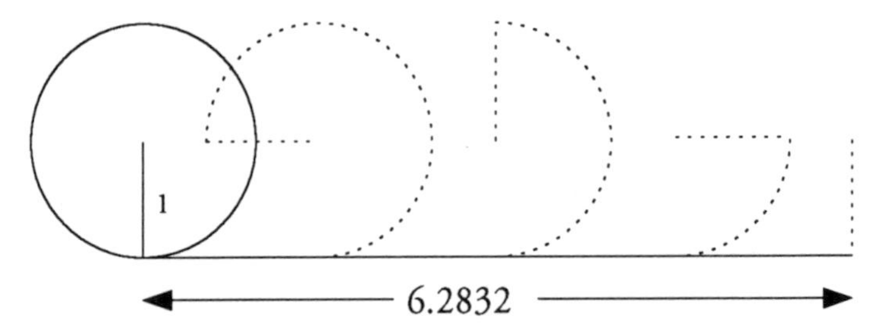

A 360° angle on the unit circle has an arc length of 6.2832 or 2π.

Why is this circumference twice as long as in the diagram on page 86
Remember: Diameter = 2 radii. The diameter of a unit circle is two.

- To find the arc length for a 1° arc in the unit circle, divide 2π by 360°.

$$1° = \frac{2\pi}{360} = \frac{\pi}{180}$$

$\frac{\pi}{180}$ is the arc length of 1° in a unit circle

- To find an arc length on the unit circle, multiply the number of degrees in the angle by the arc length for 1° ($\frac{\pi}{180}$).

For example: Find the arc length of a 70° angle on the unit circle.

$$70° \times \frac{\pi}{180} = \text{arc length}$$
$$\frac{70}{1} \times \frac{\pi}{180} = \text{arc length}$$
$$\frac{70 \times \pi}{180} = 1.222$$

Seeing is Believing: Page 196

Practice 39: Find the arc lengths for each of the below angles of a unit circle. Round your answers off to four decimal places. Answers-Page 220.

Remember: $\frac{\pi}{180}$ = arc length of 1° in a unit circle.

	Degree of angle		Degree of angle
(1)	45°	(5)	79°
(2)	10°	(6)	1°
(3)	120°	(7)	14.5°
(4)	350°	(8)	$99\frac{1}{2}°$

notes

Radians

It takes a lot of words to say, "A 1° central angle on a unit circle has an arc length of 0.017453." Mathematicians call the arc lengths of the unit circle, **radians**. They simply say that a particular angle equals so many radians, such as 1° = 0.017453 radians or $\frac{\pi}{180}$ radians.

Radians are the *"base of 1" we were seeking for arcs.*

The central angle of 1° is equal to $\frac{\pi}{180}$ radians. If you multiply a central angle (70°) by the radians for 1°, the answer is the radians for that central angle. Here is the formula:

$$\text{Angle } \theta = \theta \times \frac{\pi}{180} \text{ radians} = \frac{\theta\pi}{180} \text{ radians}$$

Remember: θ is the variable for any angle.

$$70° = 70 \times \frac{\pi}{180} \text{ radians or } \frac{70\pi}{180} \text{ radians or } 1.22173 \text{ radians}$$

Every angle has a radian measurement. If you look at the function tables on pages 214-215, you will find the radians for 0°-90°. The radian measurement beside 70° equals 1.2217.

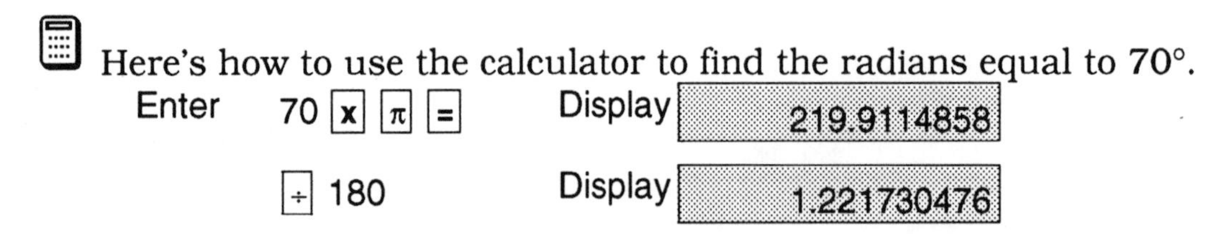

Practice 40: Use your calculator to find the radian measurement for each angle. Answers-Page 220.

(1) 20°

(2) 45°

(3) 37°

(4) 60°

(5) 1°

(6) 129°

(7) 2°

(8) 224°

Finding the Arc Length Using the Radius and Radians

The next step is to use those radians to find arc lengths. Here's the formula.

Arc Length = Radians x Radius

In the previous section, you found that the radian measurement for a 70° angle is 1.22173. To find the arc length of a 70° arc with a 10" radius, multiply the radian measurement by the radius.

Arc Length = Radians x Radius

Arc Length = 1.22173 x 10"

Arc Length = 12.2173"

Example: What is the arc length of a 43° arc with a radius of 9"?

Arc Length = Radians x Radius

Arc length = $\frac{43° \times \pi}{180}$ x 9"

Arc length = 0.7505 x 9"

Arc length = 6.75" or 6 $\frac{3}{4}$"

Enter	43 [x] [π] [=]	Display	135.0884841	
Push	[÷] 180 [=]	Display	0.750491578	43° radian
Push	[x] 9 [=]	Display	6.754424205	Arc length

Practice 41: Calculate the length of each arc using the central angle and radii measurement. Convert your answers to whole numbers and fractions. Answers-Page 220.

	Angle	Radius		Angle	Radius
(1)	61°	8"	(7)	33°	10'
(2)	54°	12"	(8)	90°	3.75"
(3)	45°	4.5"	(9)	78°	86'
(4)	90°	4"	(10)	91°	6.465"
(5)	165°	7.5"	(11)	4°	68"
(6)	6°	12'	(12)	100°	100"

Notes on Radians

You first learned that the unit of measurement for angles is degrees. Another unit of measurement that we use for angles and arcs is radians. Every angle measured in degrees also has a radian measurement. Just as the 45° angle is a 45° angle no matter how large or small the triangle or circle it is a part of, the radian measure of all 45° angles is .7854 no matter what size the triangle or circle.
45° = .7854 radians.

The advantage of using a radian measurement to describe an angle is that you can multiply the radian measurement by the radius of a circle and find the arc length of the angle. You cannot multiply the degree measurement by the radius to find arc length; however, you can convert the degree measurement to radian measurement, then multiply.

radians x radius = arc length

More Facts on Radians

- One radian is equal to the length of the radius of the same circle. The radian is a curved line and the radius is a straight line.
- There are 2π (6.28) radians in every circle.
- The measurement of any angle can be expressed in radians or degrees.
- To convert degrees to radians, multiply the degrees by $\frac{\pi}{180}$.
- To convert radians to degrees, multiply the radians by $\frac{180}{\pi}$.

Chords

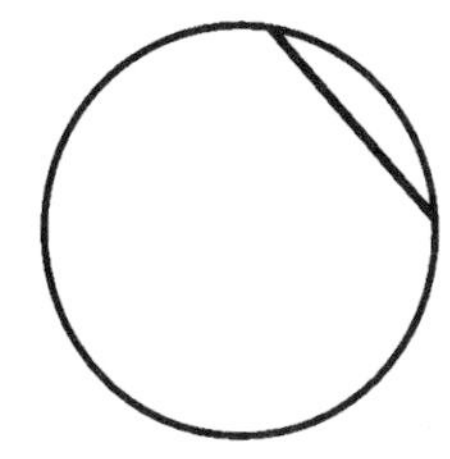

Chords are straight lines that join one point on a circle to another point on the circle.

Remember: **The diameter is the longest chord in any circle.**

Chord calculations are used with polygons, mitered turns, saddles, and many other parts of our work. Let's learn about chords using our knowledge of isosceles and right triangles. (If you feel uncomfortable with isosceles triangles, return to page 75 and rework it.)

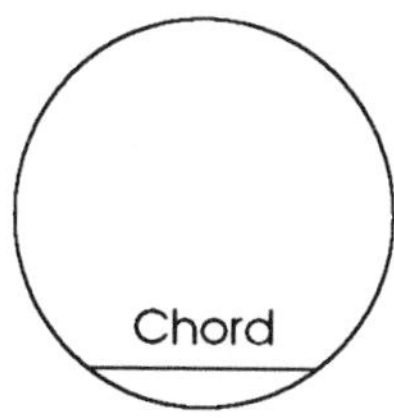

The chord is the base or unequal side of the isosceles triangle. The equal sides of the triangle are the radii that connect the ends of the chord to the center of the circle.

Finding the Angle Measurement of a Chord

In this example, the *knowns* are the radii and the chord length. The *need* is the central angle.

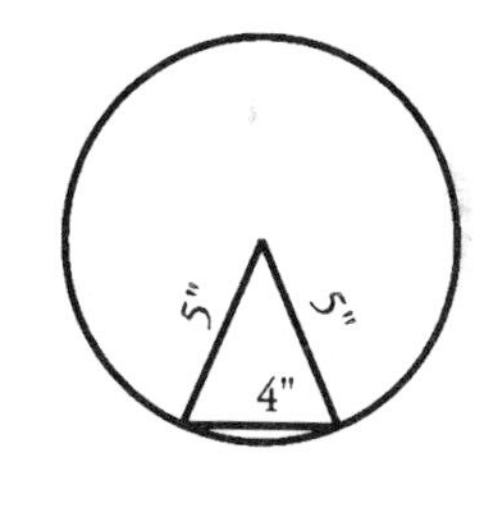

The radius of this circle is 5" and the chord length is 4". Notice that this triangle has two equal sides (5"). That means it is an isosceles triangle.

Remember: **An important property of an isosceles triangle is that a bisecting line can be drawn from the vertex of the odd angle to the center of the base and form two equal right triangles.**

Divide the isosceles triangle into two equal right triangles by dropping a bisecting line. The right triangles have two *knowns*—the length of the hypotenuse and the length of the opposite side. The reference angle is the central angle.

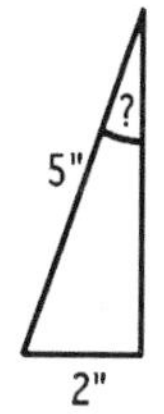

The length of the hypotenuse of the right triangle is the radius of the circle (5"), and the length of the opposite side is half of the chord (2").

The function chart shows that there are two functions which can be used if you know the length of the hypotenuse and the length of the opposite side.

Sine θ	=	$\frac{\text{Opposite side}}{\text{Hypotenuse}}$	Cosecant θ	=	$\frac{\text{Hypotenuse}}{\text{Opposite side}}$
Cosine θ	=	$\frac{\text{Adjacent side}}{\text{Hypotenuse}}$	Secant θ	=	$\frac{\text{Hypotenuse}}{\text{Adjacent side}}$
Tangent θ	=	$\frac{\text{Opposite side}}{\text{Adjacent side}}$	Cotangent θ	=	$\frac{\text{Adjacent side}}{\text{opposite side}}$

This time we will use **sine** function.

Sine θ = $\frac{\text{opposite side}}{\text{hypotenuse}}$

Sine θ = $\frac{2}{5}$

Sine θ = .4

Sine **23.6°** = .4 This can be done with the calculator or the function chart in the back.

Round off to $23\frac{1}{2}°$.

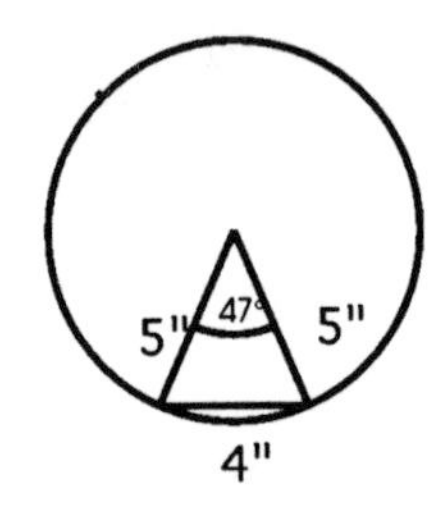

Since there are two equal right triangles in the isosceles triangle, you must double this angle to get the central angle of the chord.

$2 \times 23\frac{1}{2}° = 47°$

Practice 42: Find the central angle of each chord. Round your answers off to the nearest tenth degree. Answers-Page 220.

	radii	chord length		radii	chord length
(1)	3"	1"	(5)	12"	1' 6"
(2)	6'	6"	(6)	62"	13"
(3)	1.333'	.333'	(7)	5.5'	2.75'
(4)	92'	76"	(8)	5.5"	10.75"

Chord Length

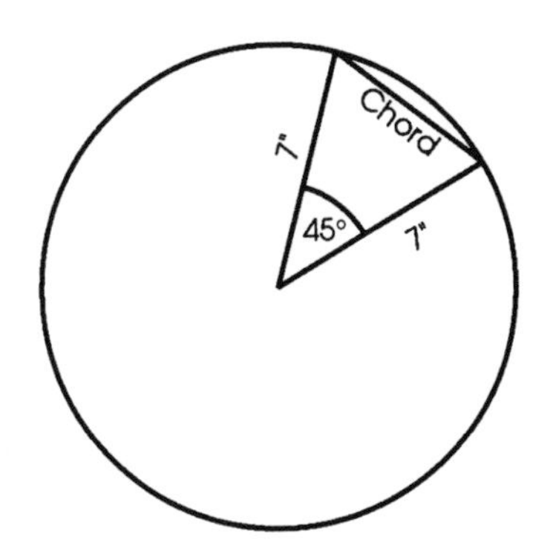

The *knowns* of this chord are the central angle (45°) and the radius of the circle (7").
The *need* is the chord length.

Divide the isosceles triangle into two right triangles and state your *knowns* and *need.*

The *knowns* are one angle and the length of the hypotenuse.
The *need* is the length of the opposite side of the known angle.

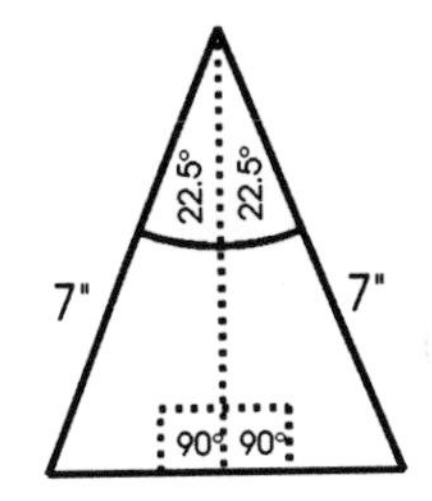

Hypotenuse	=	csc θ	x	Opposite
Hypotenuse	=	sec θ	x	Adjacent
Opposite	=	tan θ	x	Adjacent
Opposite	**=**	**sin θ**	**x**	**Hypotenuse**
Adjacent	=	cos θ	x	Hypotenuse
Adjacent	=	cot θ	x	Opposite

Opposite side = sin 22.5° x 7"

sine 22.5° = 0.3827

Opposite side = 0.3827 x 7"
Opposite side = 2.679" or $2\frac{11}{16}$"

Remember: This measurement is only half the length of the chord. To find the length of the chord, double it.

4.9497" x 2 = 9.8994" or $9\frac{7}{8}$"

Practice 43: Find the length of each chord. Convert your answers to whole numbers and fractions. Answers-Page 220.

	central angle	radius		central angle	radius
(1)	45°	45"	(4)	120°	2"
(2)	90°	10'	(5)	37°	14' 6"
(3)	15°	3' $4\frac{9}{16}$"	(6)	60°	30'

notes

Second Module Preface

All math has a practical use: It is a method to solve problems. To solve problems with the math you have learned, you must know how it applies to the work you do. The next module shows ways to make the math you've learned practical.

Craftsmen of all types have been writing down methods of doing tasks for hundreds of years. The applications in this next module are derived from many past experiences with applying math to work problems. In the first module, you learned to find lengths of the straight lines of right triangles and lengths of the curved lines of circles. Placing material in a straight line is not difficult; however, turning an exact angle or using several turns to move to an exact location is more difficult. The next module is a study of turns, offsets, and other geometric patterns that are useful to the crafts people in their work.

The information presented in Module 2, like the information in the rest of the book, is fairly general in nature. There may be parts that you think are not used in your trade, and you may be correct, but that doesn't mean they can't be. If you keep an open mind to their uses, these applications can give you new approaches to solving problems in your field of work. In the first module, a stair step learning pattern was used. Each step you learned applied to the following steps. In the second module, you will be on a learning ladder. You will learn a series of skills that do not appear related, but in the end they will tie together. This process is not much different than the way you do your work in the field. You don't stop work because you can't do a particular part of a job; you do another part that will fit in later. When you combine all the parts, the task is complete.

notes

XII

Offsets

Pieces of the Puzzle and Invisible Lines

An offset is generally used to get around an obstacle or to reach a new elevation. **To offset means to move over, up, or down to reach a new path going in the same direction.**

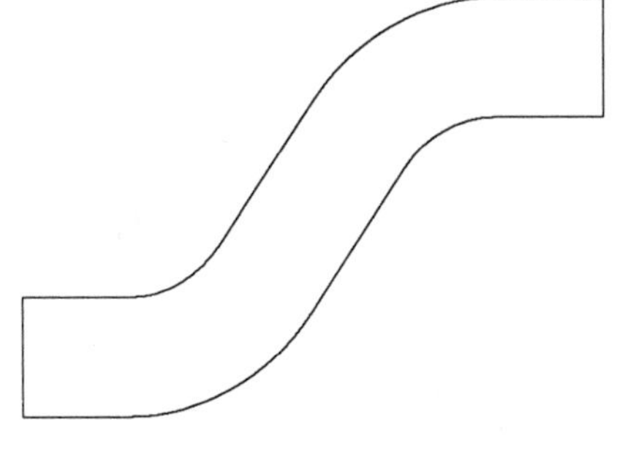

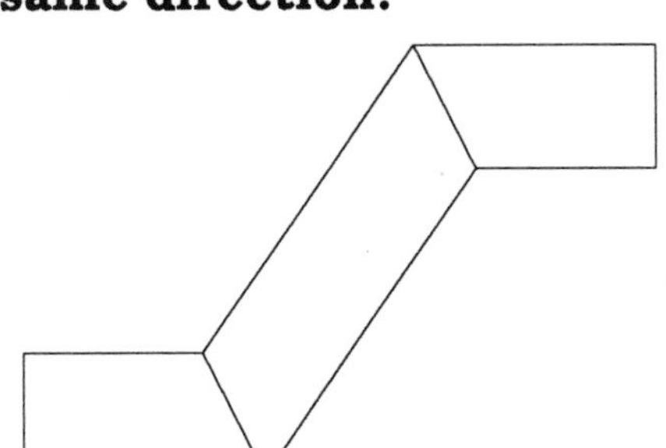

These two drawings show almost all you will see when you look at an offset in the field. Both make the same offset. One does it with miters, the other uses arcs.

Below you can see the center lines of both offsets. Notice the center lines for both offsets are parallel lines crossed by a straight line. You can refresh your memory about straight lines crossing parallel lines by referring back to page 32.

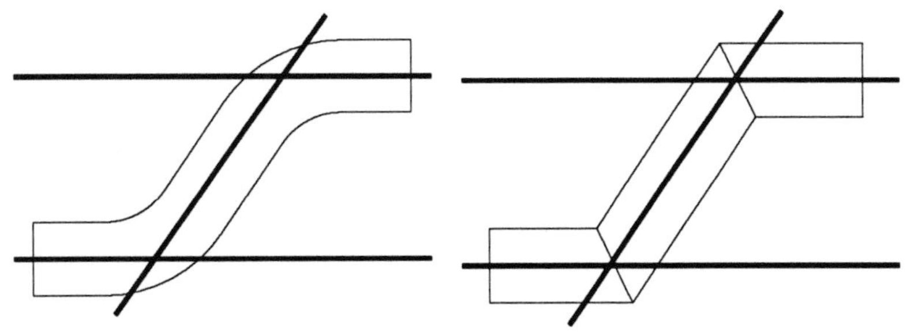

Center lines, like many of the lines we work with in offsets, are invisible. We draw sketches to illustrate where these lines are and to record measurements. You will be encouraged in the next sections to draw thumbnail sketches and should find them helpful not only in this book, but also in the field.

The drawings below show the invisible lines that you will learn to use when calculating offsets.

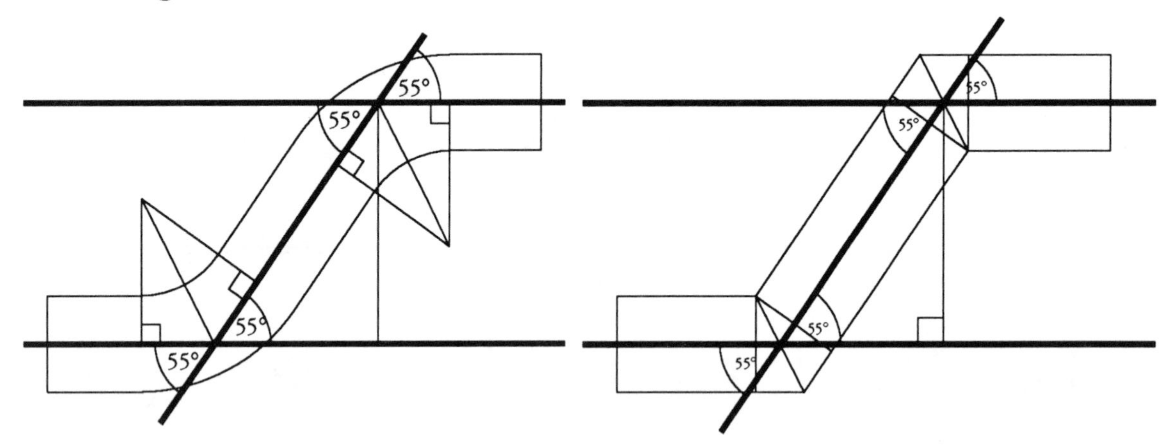

These drawing look fairly complicated. They certainly are complicated if you try to think about all of the information in them at once. The trick is to work one piece at a time. First draw a thumbnail sketch, then, bit by bit, do the calculations. When you finish with one set of calculations, add those numbers to your sketch and go on to the next part. At the end, you will have all the parts and the numbers to make a complete offset.

Simple Offsets

Remember: To offset means to move over, up, or down to reach a new path going in the same direction.

A *simple offset* uses two turns of the same angle to reach that new path. One turn changes the direction and the other turn renews the original direction.

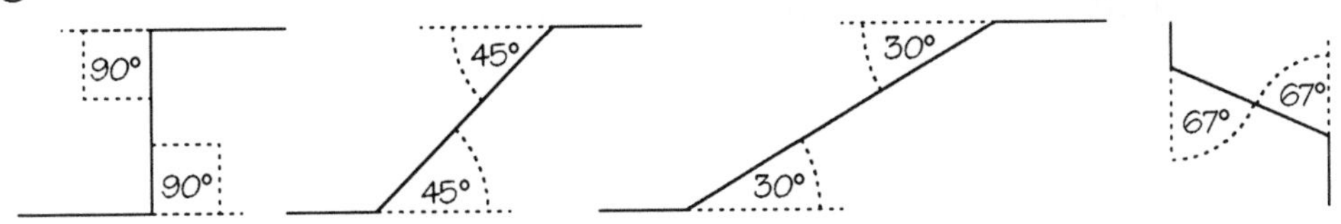

Planes and Offset Boxes

With simple offsets, the center line of all pieces are generally in either a horizontal (level) plane or a vertical (plumb) plane.

- If you want to *change elevation*, you make a **vertical simple offset**.
- If you want to *offset to the side* (on the same elevation), youmake a **horizontal simple offset**.

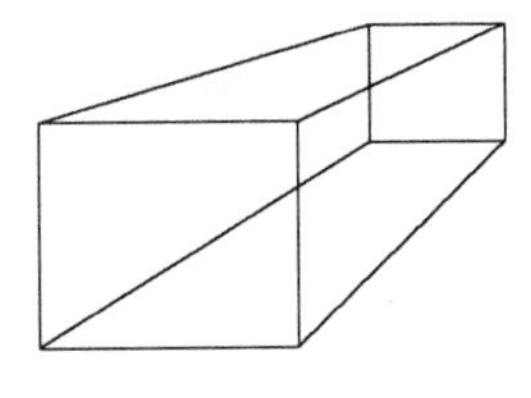

Many crafts people use an imaginary box when working offsets. Since the sides of a box are vertical planes and the top and bottom are horizontal planes, the box can be used to show any simple offset.

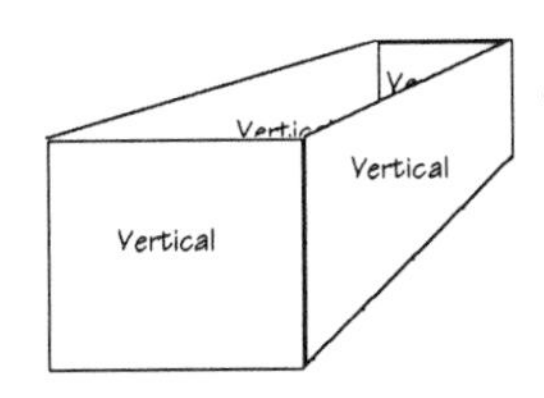

The four sides of a box are vertical planes. *The vertical simple offsets can be drawn on any one of the four vertical sides.*

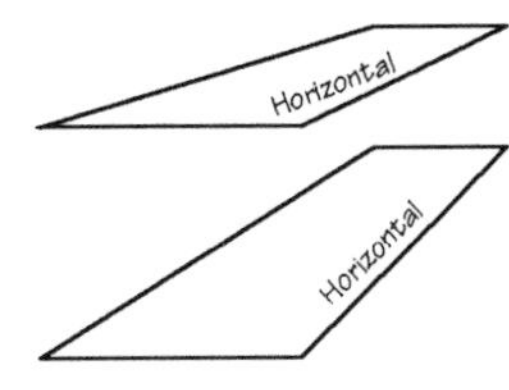

The top and bottom sides are horizontal planes.

Notice the drawing on the next page.

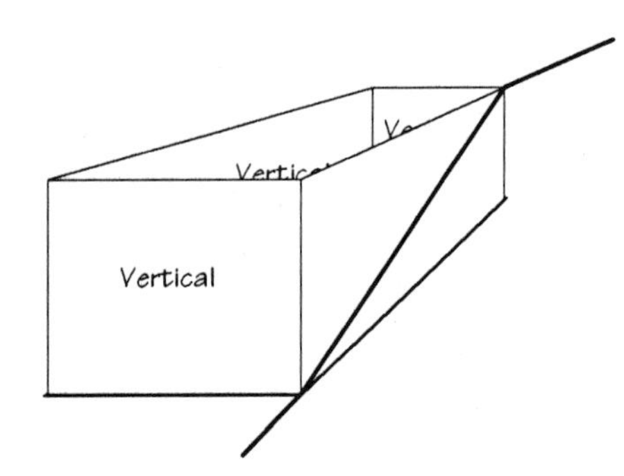

Remember: **If you want to change the elevation of a line then correct back to the same direction, use a *vertical simple offset.***

Horizontal simple offsets can be drawn on the horizontal sides (the top or bottom) *of the box. Notice* the drawing below.

Remember: **If you want to move a line over on the same elevation, use a *horizontal simple offset.***

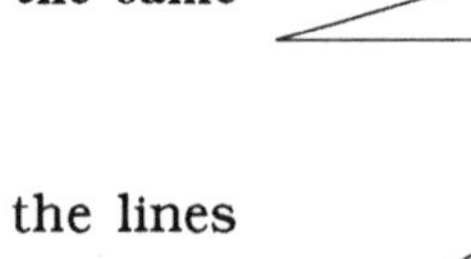

NOTE: The perspective of the drawings above makes the lines look as if they are not parallel; however, they are.

Now that you have seen how simple offsets are drawn on the sides of an offset box, we can work with these offsets using only the rectangle (or side of the box) that the offset was drawn on. If you look directly at a side on which an offset has been drawn, you will see a rectangle (or square) with a diagonal line drawn through it.

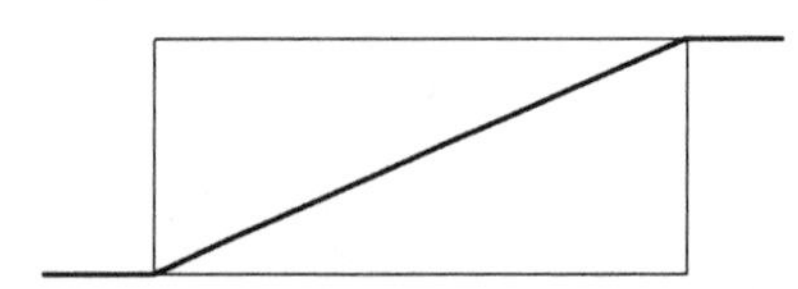

Any simple offset, vertical or horizontal, can be drawn using a rectangle or square. The following drawings show examples of different offsets.

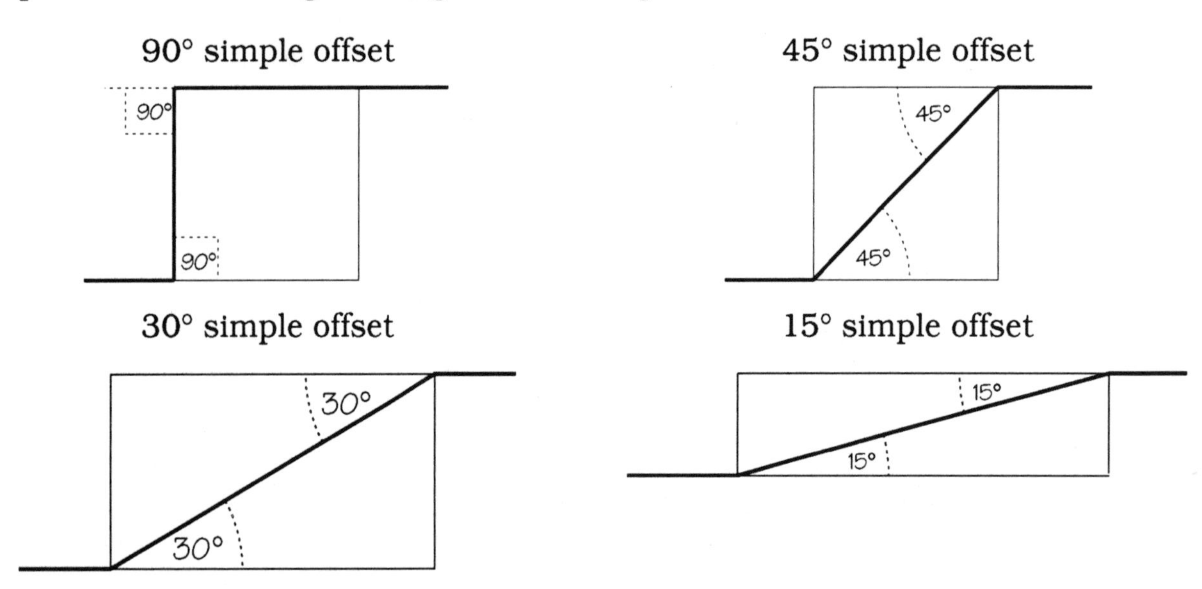

With all simple offsets except the 90° simple offset, a diagonal line bisects a rectangle, creating two equal right triangles. *All offsets except the 90° offset use right triangles to calculate distances.* These right triangles are called **offset triangles** and are used in the solving of simple offsets.

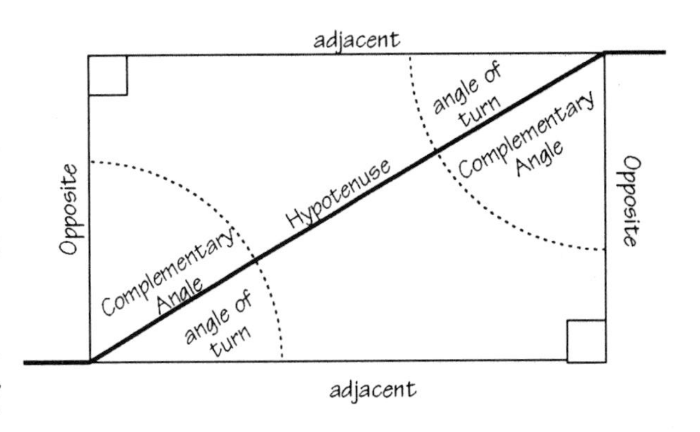

Remember: **A diagonal always divides a rectangle exactly in half and creates two equal right triangles.**

The information you find for one triangle is the same for the other. The diagonal line which divides the rectangle into two equal right triangles is the **hypotenuse** for both triangles.

Notice there is one angle of turn in each of the right triangles. *The angle of turn is also the angle of the arc, elbow, or bend to be used for the offset.*

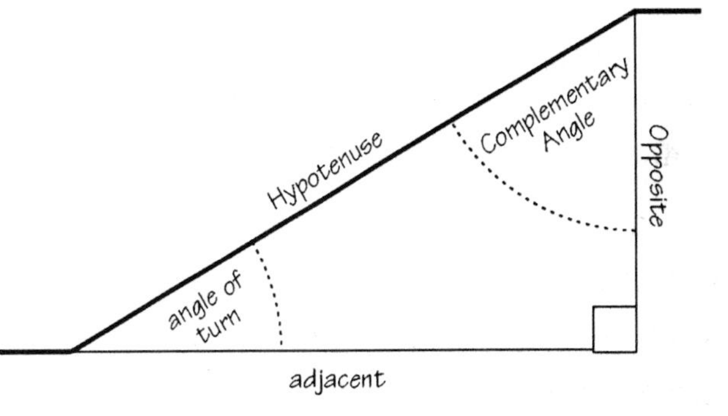

notes

notes

Angle of Turn

It is important to know where the **angle of turn** is located. Many assume that this angle is an interior (inside) angle. Instead, it is *the angle behind the turn.*

The angle of turn is an angle formed between where the original line of travel would have taken you (if you had not turned) **to the new direction of travel.**

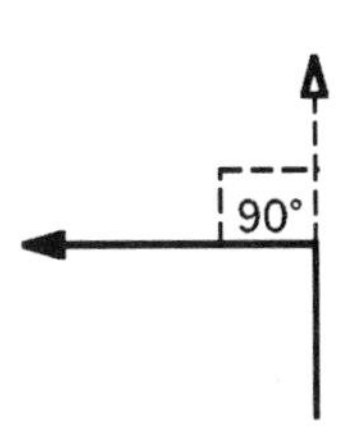

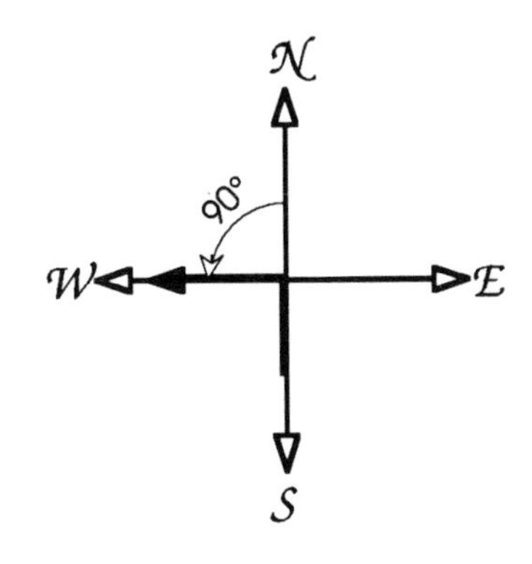

Think about this in terms of the compass. The major directions of a compass are north, east, south, and west. Each major compass point is 90° from the one next to it. If you are traveling north and turn west, you change your direction of travel by 90°. The angle of turn is measured between the north and west, not the south and west.

Between the major compass points are other directional arrows which are 45° from the major ones. Their names are a combination of the nearest major directions—northeast, southeast, southwest, and northwest. If you are traveling east and turn northeast, the angle of turn is 45° and is measured between east and northeast.

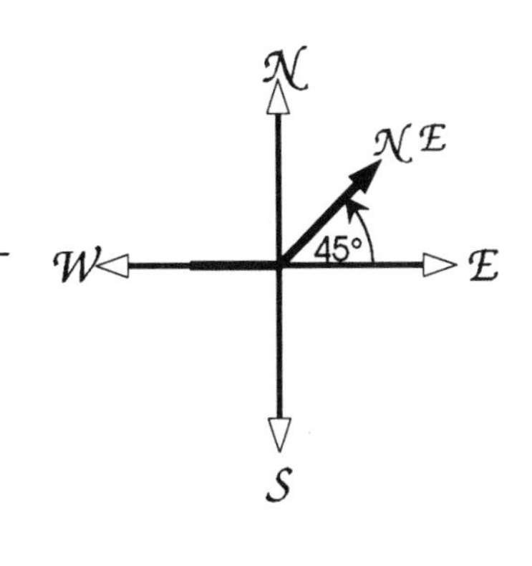

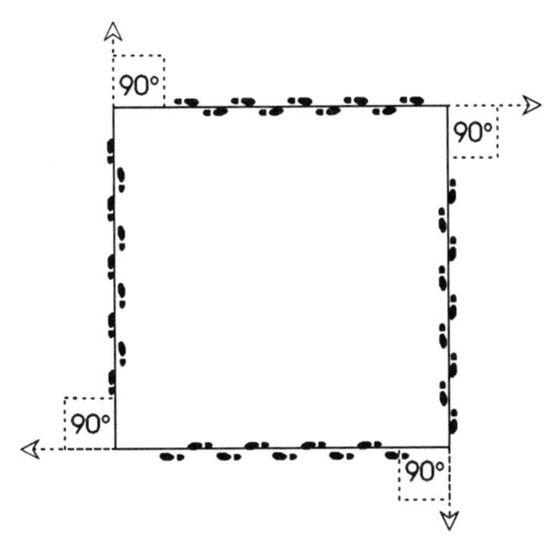

Corners have two angles of turn. The angle of turn you use depends on the direction in which you travel. This drawing shows the angles of turn made if you were to walk one way around a city block.

If you turned around and walked the same block in the opposite direction, the angles of turn would change. This is shown in the drawing below.

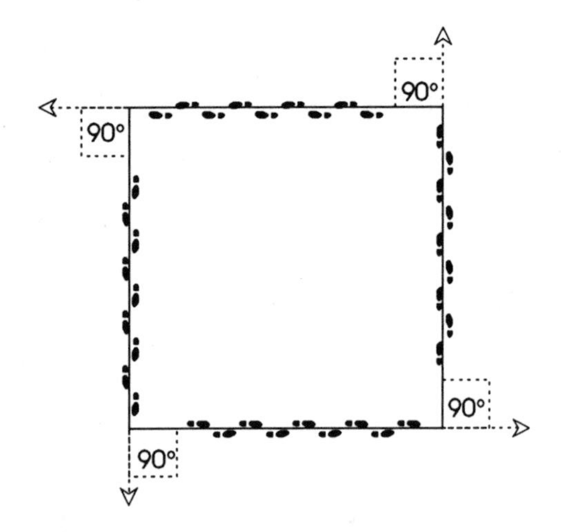

Notice at each corner there are two possible angles of turn. They are opposite angles which are always equal. The one you use depends on your direction of travel.

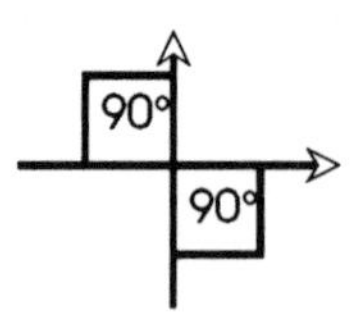

Remember: **The angle of turn is always behind the turn.**

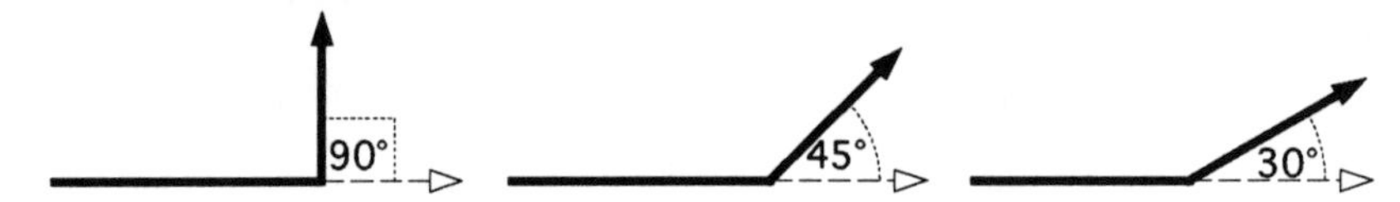

Remember: **The angle of turn can be drawn for either direction.**

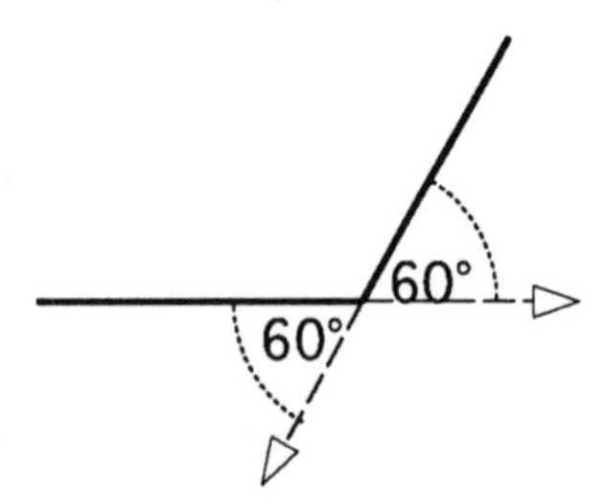

Offset Triangles

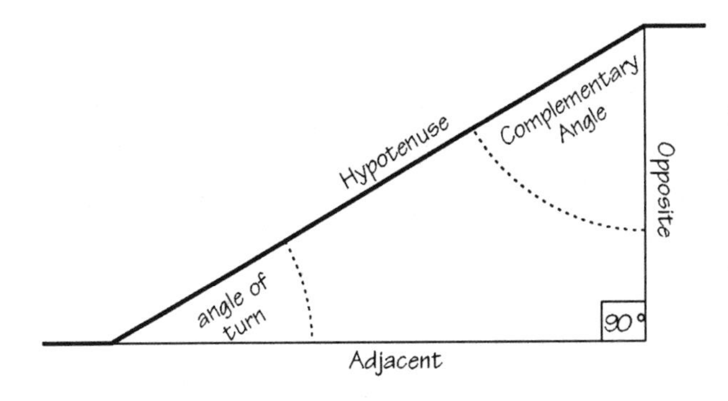

Many trades use special names when referring to the sides of an offset triangle. We will continue to use the names of the sides of the right triangle in order to maintain the triangle's relationship with the functions and the NAK chart.

With an offset, the angle of turn is generally considered the reference angle and the known side is the opposite side.

The opposite side is the distance between the parallel lines of an offset.

Simple Offset Triangles

90° Simple Offsets

Remember: **All offsets except the 90° offset use right triangles to calculate distances.**

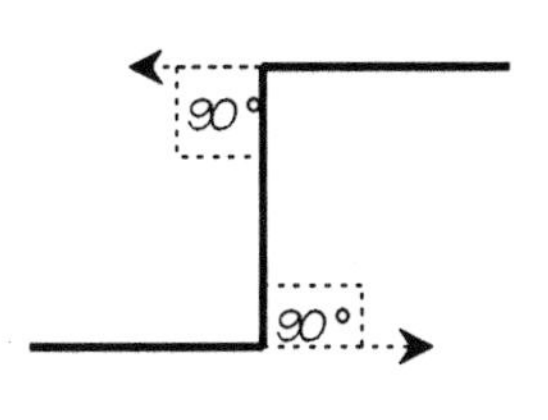

Most offsets are made using 90° turns. This is a very easy offset to calculate since the length of the offset is the same measurement as the distance you want to move over, up, or down.

Example: If the distance you need to move is 12", then the length of the offset is 12".

- There is no practice needed for the 90° simple offset since no offset triangles are involved. The length of the offset is the same as the offset amount.

45° Simple Offsets

The 45° simple offset is often used because the 45° right triangle has two equal legs. The *adjacent* and *opposite sides* are equal.

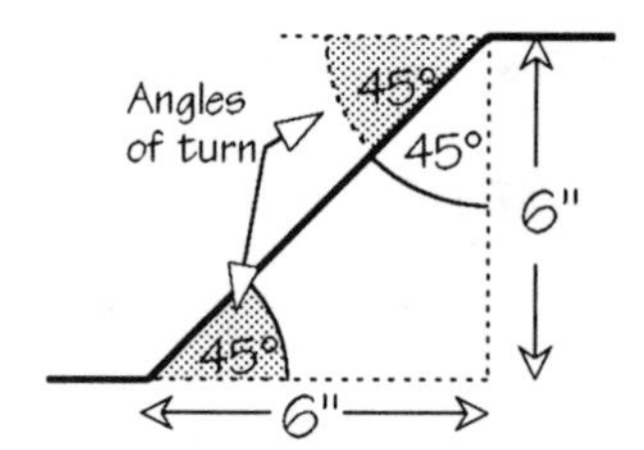

Remember: To find the length of the hypotenuse when the length of one of the legs of a 45° right triangle is known, multiply the leg by 1.4142 or $\sqrt{2}$. **(page 79).**

Example: Find the length of the hypotenuse of the 6" offset above.

Hypotenuse = opposite side x 1.4142

Hypotenuse = 6" x 1.4142

Hypotenuse = **8.49"**

Practice 44: Find the length of the hypotenuse for each 45° simple offset. Convert your answers to whole numbers and fractions. Answers-Page 220.

Remember: The lengths of the opposite and adjacent sides are equal in 45° simple offsets.

	Opposite side or Adjacent side		Opposite side or Adjacent side
(1)	10"	(5)	21.125"
(2)	22'	(6)	78"
(3)	7.375"	(7)	14.875"
(4)	43"	(8)	57.0625"

30° Simple Offsets

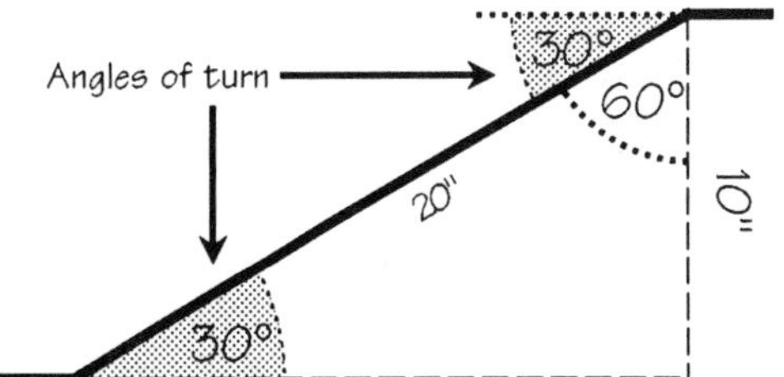

The 30° simple offset is another easy offset to calculate. Electricians prefer the 30° simple offset over the 90° or 45° because wire pulls more easily through conduit with smaller angle bends.

Remember: The length of the hypotenuse is always twice the length of the shortest side in a 30°- 60° right triangle.

If you want to offset 10", then the length of the hypotenuse is 20" (See page 81 if you do not understand why).

To find the length of the adjacent side, use the NAK chart.

NAK Chart				
Need		Angle		Known
Hyp	=	csc θ	x	Opp
Hyp	=	sec θ	x	Adj
Opp	=	tan θ	x	Adj
Opp	=	sin θ	x	Hyp
Adj	=	cos θ	x	Hyp
Adj	=	cot θ	x	Opp

The *knowns* are the length of the opposite side and a 30° angle.
The *need* is the length of the adjacent side.

Adj. = cot θ x Opp.
Adj. = 1.732 x 10"
Adj. = **17.32"**

Practice 45: Find the length of the adjacent side and the length of the hypotenuse for each 30° simple offset. Round your answers off to four decimal places. Answers-Page 220.

	Opposite side		Opposite side
(1)	75.25"	(5)	45'
(2)	13.5625"	(6)	26.125"
(3)	7'6"	(7)	4'
(4)	3"	(8)	11.25'

The quickest way to enter 1.732 in your calculator is to enter 3, then push $\boxed{\sqrt{x}}$.

notes

Odd Angle Simple Offsets

Odd angle simple offsets are offsets other than the 90°, 45°, and 30° simple offsets. They are generally used in areas with limited space. Instead of deciding to use one of the easier angle of turn (such as 90°, 45°, or 30°), you must calculate the angle of turn based on the room you have for an offset.

Example: In order to bypass two obstacles, you need to move up 23" and over 85.837". Find the angle of turn and the length of the hypotenuse for this odd angle simple offset.

First: Draw a thumbnail sketch and write in your *knowns*. State your *needs*.

- The *needs* are the measure of the angle of turn *and* the length of the hypotenuse.

Second: Ask yourself, "What do I know that uses these *knowns* to find the *needs*?"

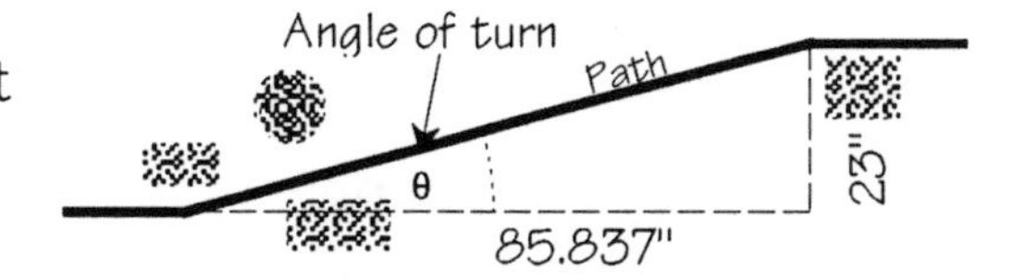

- The functions chart shows that the formula needed to calculate this angle of turn is: Tangent $\theta = \frac{\text{opposite side}}{\text{adjacent side}}$.
- By using the Pythagorean theorem you can determine the length of a third side of a right triangle when two sides are known.

Remember: The angle of turn is the reference angle. It is also the angle of the elbow or bend to be used.

Third: Calculate the problems.

- Find the angle of this right triangle by using the tangent function.

$$\text{Tangent } \theta = \frac{\text{opposite side}}{\text{adjacent side}}$$

$$\text{Tangent } \theta = \frac{23}{85.837}$$

$$\text{Tangent } \theta = 0.26795$$

$$\text{Tangent } 15° = 0.26795$$

$$\text{Angle of turn} = \mathbf{15°}$$

- Find the length of the hypotenuse of this 15° simple offset.

  Your *knowns* are the length of the adjacent and opposite sides (23" and 85.837") *and* a 15° angle.

Notice that since you now know a smaller angle as well as two sides of the right triangle, you can find the third side by using either the Pythagorean Theorem *or* the NAK chart.

$$c = \sqrt{a^2 + b^2}$$
$$c = \sqrt{23^2 + 85.837^2}$$
$$c = \sqrt{529 + 7368}$$
$$c = \sqrt{7897}$$
$$c = 88.865$$

Hypotenuse = **88.865"**

Practice 46: Find the angle of turn and the length of the hypotenuse for these odd angle offsets. Round each answer off to the nearest .5°. Answers-Page 220.

	Opposite side	Adjacent side		Opposite side	Adjacent side
(1)	10"	17"	(5)	$33\frac{1}{2}$"	$12\frac{9}{16}$"
(2)	5' $7\frac{3}{4}$"	2'1"	(6)	22'	53'
(3)	43"	28"	(7)	$4\frac{3}{4}$"	6"
(4)	7' $6\frac{5}{8}$"	3' 5"	(8)	3'	4'

Offset Turns

Offset turns use more than one angle to *change direction*. When the location where you normally would make a 90° turn is blocked, two or more turns may be combined to make a total turn.

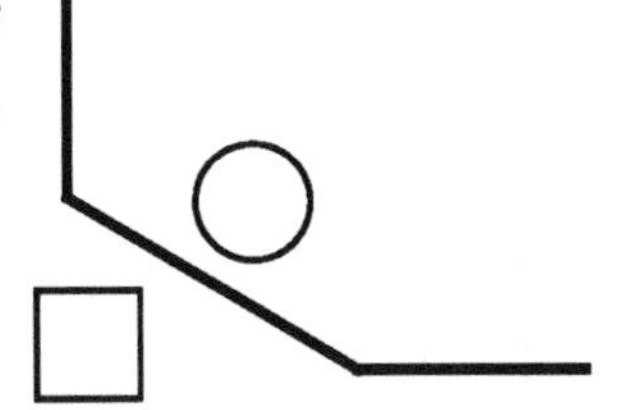

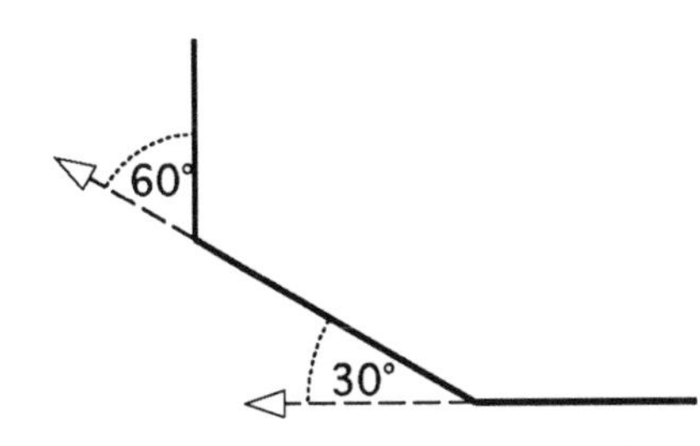

In these drawings, a total turn of 90° is being made by two separate turns—one of 30° and one of 60°. *The degrees in the individual turns of an offset turn must equal the degrees in the total turn.*

Notice that the offset triangle for this offset turn is a 30°- 60° right triangle. *The offset triangle for an offset turn has two angles of turn.* This offset triangle has a 30° turn and a 60° turn. The length of the hypotenuse is calculated the same way as with a simple offset.

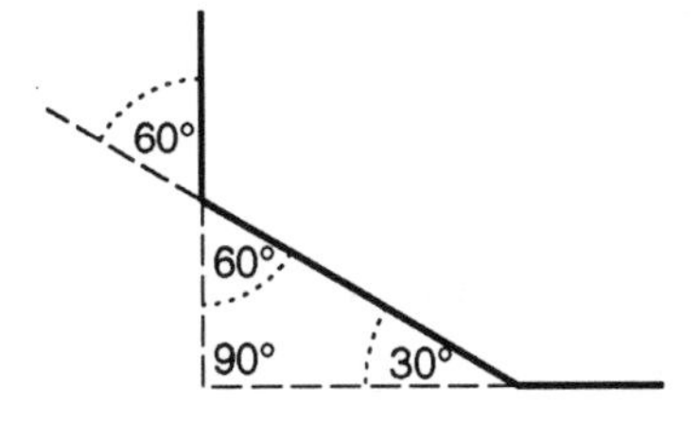

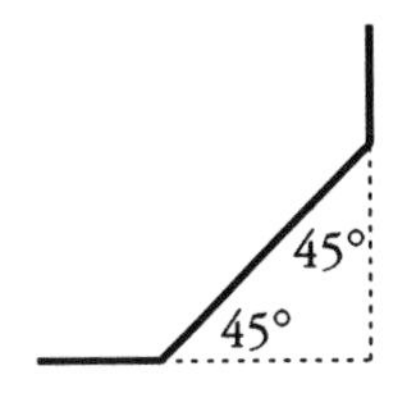

The 45° offset turn is the most often used offset turn. It is easy to calculate since the two legs are equal and both angles of turn are the same (45°). The turn uses the same offset triangle as the 45° simple offset. The difference is that there are two angles of turn in this offset triangle.

Practice 47: Find the angles of the turn and the length of the hypotenuse for each offset turn. Round each angle off to the nearest .5° and the decimal numbers to four decimal places. Answers-Page 220.

	Opposite side	Adjacent side		Opposite side	Adjacent side
(1)	24"	14"	(5)	4'	4'
(2)	63"	54"	(6)	38"	14"
(3)	12'	12'	(7)	51"	88.33"
(4)	102'	50"	(8)	42.56"	32.25"

notes

Concentric Arcs

Since the materials we work with have width, we may have to consider more than one arc in solving some problems. Generally there are three radii and their arcs: inside, center, and outside. The radius given on blueprints is usually the center radius (C.R.) of the arcs.

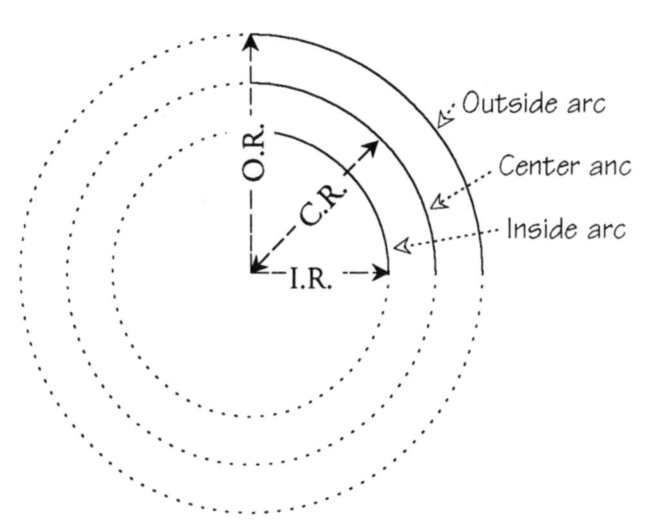

Remember: **Concentric arcs have the same center point but different radii lengths.**

(See page 93 if you have difficulty with concentric arcs.)

There is a step by step process for finding the arc lengths of concentric arcs.

First: Find the radian measurement of the angle of the arcs.

Second: Find the radius measurement of each arc. The center radius is usually given. The inside and outside radii are found by dividing the width of the material being used by two, then subtracting that answer from the center radius for the inside radius (I.R.) and adding that answer to the center radius for the outside radius (O.R.).

Third: Multiply the radian measurement by each of the radii.

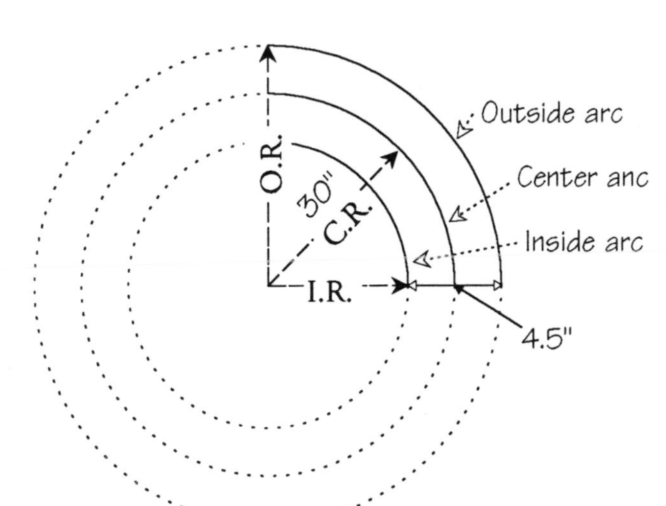

For example: A hydraulic bender for 4" rigid conduit has a center radius of 30 inches. The outside diameter of 4" rigid conduit is 4.5". Find the three arc lengths of a 30° bend.

(1) Radians $30° = \frac{30 \times \pi}{180}$ radians = **0.5236 radians**

Remember: $1° = \frac{\pi}{180}$ radians (See page 96)

(2) Radii Outside diameter = 4.5" (given)

$\frac{1}{2}$ O.D. = $\frac{4.5}{2}$ or 2.25"

Center radius = **30"**

Inside radius = 30" - 2.25"= **27.75"** (Center Radius - $\frac{1}{2}$ O.D.)

Outside radius = 30" + 2.25"= **32.25"** (Center radius + $\frac{1}{2}$ O.D.)

(3) Arc Length **Radians (30°=*0.5236* radians) x radius = Arc length**

(See page 96 if you have difficulty with radians)

Inside arc = *0.5236* x 27.75" = **14.53"**

Center arc = *0.5236* x 30" = **15.708"**

Outside arc = *0.5236* x 32.25" = **16.886"**

This may seem like a lot of calculations; however, you seldom need to calculate all of them.

- Electricians calculate mainly by using the *center arc length* which they call the *developed length of a bend.* Think about what happens to a pipe when it is bent. The outside of the conduit is stretched and the inside is compressed, so the center arc is the only one that remains the same.
- Pipe fitters, sheet metal craftsmen, and carpenters seldom use the center arc length, but they do use the inside and outside arc lengths when marking elbows for cutting. If you are forming a turn out of sheet metal or concrete, the lengths you need are the inside and outside arc lengths.

Practice 48: Find the inside, center, and outside arc lengths for each concentric arc. Round your answers off to four decimal places. Answers-Page 220.

	Center radius	Width of material	Angle		Center radius	Width of material	Angle
(1)	10"	3.5"	40°	(5)	5"	.5625"	49°
(2)	21'	6.5'	15°	(6)	6'	4'	10°
(3)	27"	14"	28°	(7)	36"	24"	90°
(4)	6"	4.5"	52°	(8)	200'	50'	45°

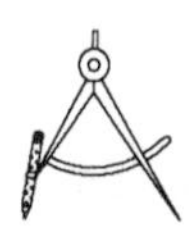

Seeing is Believing: Page 198

The Take Out or Take Up of Arcs

To use arcs in offsets and turns, such as elbows, bends, and curved forms, you must know where the arcs fit in the turns. Also, to do calculations, you must know how much room they take up in the offset triangle.

Below is a drawing of a simple offset using elbows. This drawing shows the offset triangle with the elbows in place.

The solid lines in the drawing are referred to as the *take outs* or *take ups* of the arcs. Notice that they continue the straight centerline while the arc curves around to meet the new center line.

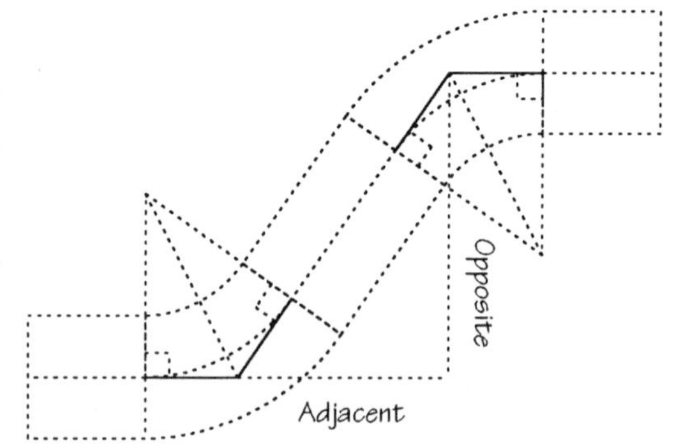

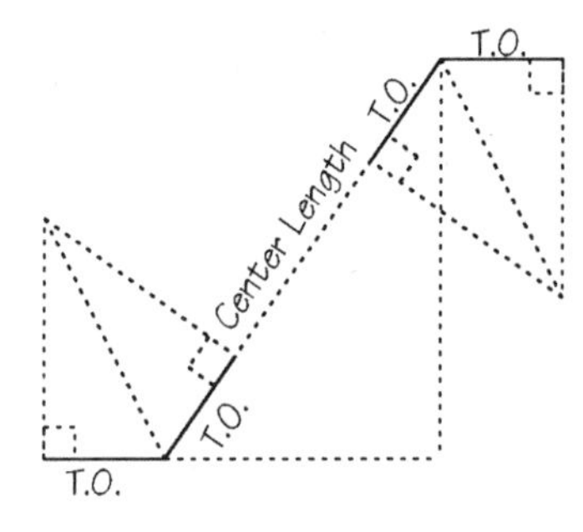

The bends or elbows are connected by a straight piece of material. This length of this straight piece of material is referred to as the **center length**.

To determine the center length, you must know how much of the length of the hypotenuse of the offset triangle is taken up by the elbows or bends. The distance that is taken up is referred to as the **take out** or **take up**.

To understand *take out or up*, we need to know about straight lines that touch just **one point** on a circle. These are called **tangent lines,** and the point where the circle and the tangent line touch is called the **point of tangency**.

If the radius of the circle is drawn so that it meets the **point of tangency**, the radius is perpendicular to the tangent line.

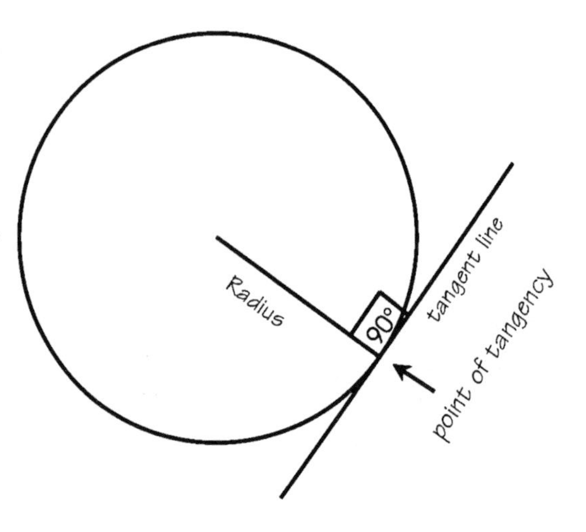

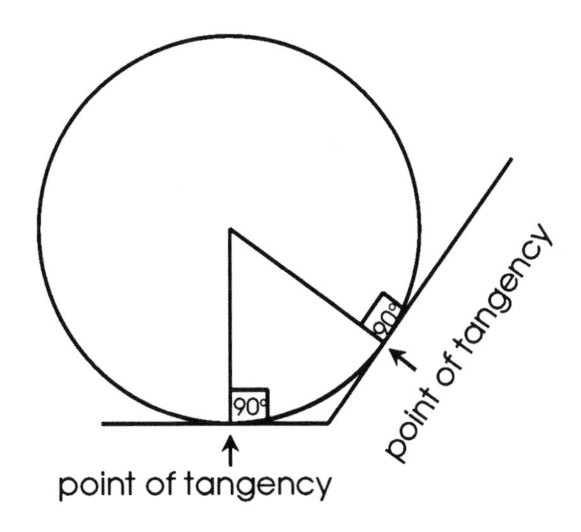

The drawing to the left shows two tangent lines meeting to form an angle. The drawing looks like a ball rolled into a corner. You can still see the corner, but the ball is as close as it can get.

Notice the angle of turn and the angle of the arc shown in the drawing to the right.

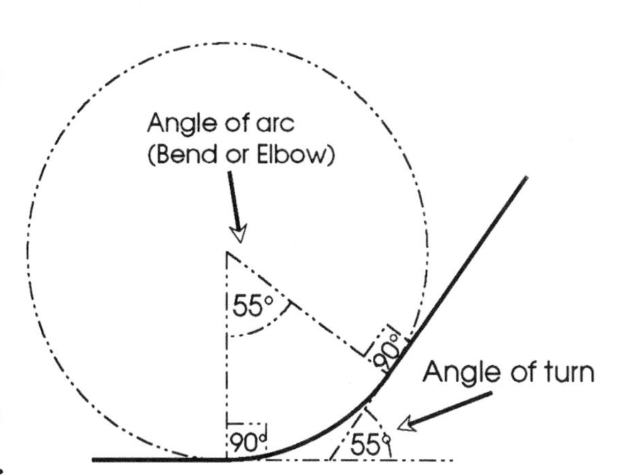

Note the following:

- **The angle of the arc is equal to the angle of turn.**
- **The radii of the arc are perpendicular to the tangent lines at the points of tangency.**
- **The transition from straight line to curved line takes place at the point of tangency.**

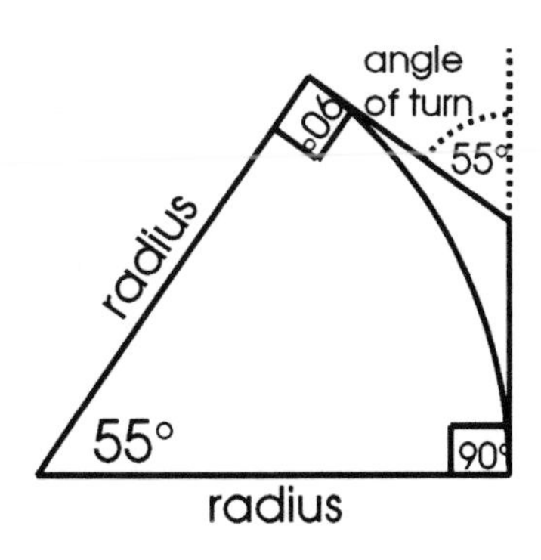

Let's take this sector showing the angle of turn out of the circle and examine it. First, look at the sector in the drawing above to make sure you know where this sector is located in the circle.

This sector looks somewhat like a kite. We are going to make it look even more like a kite by drawing a line from the vertex of the sector to the vertex of the angle created by the tangent lines. This bisecting line creates two equal right triangles.

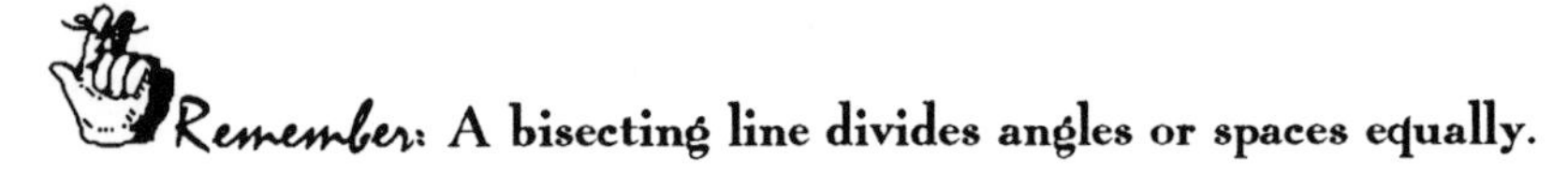

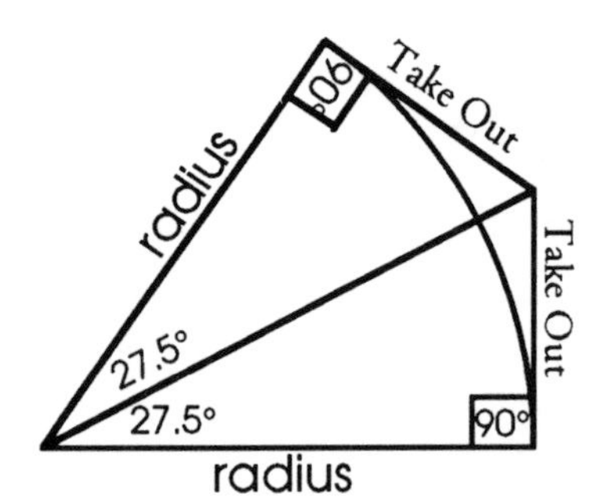

In each right triangle there are two *knowns*:

- the *angle* $27.5°$ $\left(\frac{\text{angle of turn or arc}}{2}\right)$ and
- the length of the adjacent side (radius of arc).

The *need* is the length of the opposite side (take out).

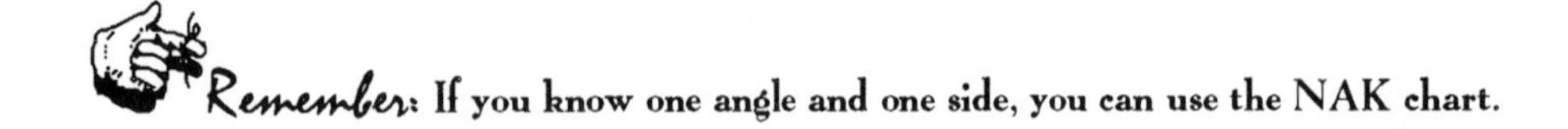

If you look at the NAK chart, you find that the needed formula is:

Opposite side = tangent θ x adjacent side.

NAK Chart				
Need		**Angle**		**Known**
Hyp	=	csc θ	x	Opp
Hyp	=	sec θ	x	Adj
Opp	=	tan θ	x	Adj
Opp	=	sin θ	x	Hyp
Adj	=	cos θ	x	Hyp
Adj	=	cot θ	x	Opp

Let's look at this equation in terms of our *needs* and *knowns*.

- *The length of the opposite side* is the take out.
- The *angle of turn* is divided by two which produces $\frac{\theta}{2}$.
- *The length of the adjacent side* is the radius of turn.

This is the working formula for take outs:

Take out = $\tan\frac{\theta}{2}$ x radius of the turn

Example: Let's say that the radius of turn is 12" and we want to calculate the take out using the working formula.

Take out = $\tan\frac{\theta}{2}$ x radius of the turn.

Take out = $\tan\frac{55°}{2}$ x 12"

Take out = tan 27.5° x 12" (tan 27.5° = 0.5206)

Take out = 0.5206 x 12"

Take out = **6.247"**

This is how the same problem is worked using the calculator.

Enter 55 [÷] 2 [=]	Display [27.5]
Push [tan]	Display [0.52056705]
Push [x] 12 [=]	Display [6.246804606]

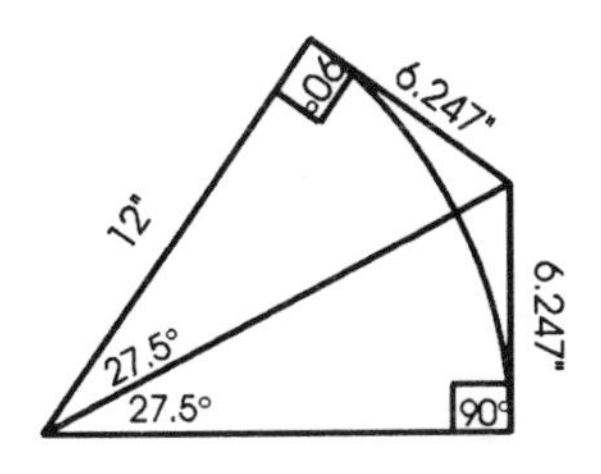

The take up or take out for a 55° elbow or bend that has a 12" center radius is 6.25".

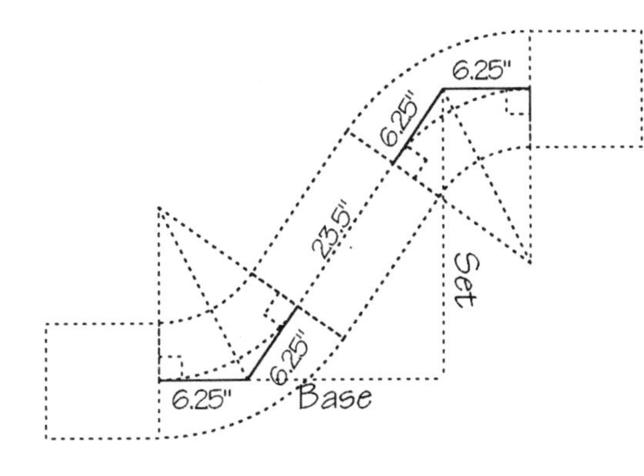

Notice in this drawing that there are two take outs on the center line (hypotenuse). Since the take outs of the elbow or bend are known, you can subtract them from the length of the hypotenuse of the offset triangle to find the center length.

The formula to find the center length is:

Center length = Hypotenuse - 2 T.O.

Example: If the length of the hypotenuse is 36", the center length is found this way:

Center length = Hypotenuse - 2 T.O.

Center length = 36" - 2 x 6.25"

Center length = 36' - 12.5"

Center length = 23.5"

Practice 49: Find the take out and the length of the center piece. Round the answers off to four decimal places. Answers-Page 221.

	Angle	Radius	Hypotenuse		Angle	Radius	Hypotenuse
(1)	33°	18"	72"	(5)	57°	36"	12'
(2)	24°	12"	42"	(6)	14°	6"	16.5"
(3)	75°	25'	150'	(7)	68°	8"	4' 11"
(4)	45°	9"	18"	(8)	90°	198'	0.5 miles*

Remember: Take out = $\tan \frac{\theta}{2}$ x radius of the turn.

Remember: Center length = Hypotenuse - 2 T.O.

notes

---

* One mile is equal to 5280 feet or 1760 yds.

notes

Simple Offsets with Arcs

90° Simple Offsets

You have already learned that for 90° simple offsets the length of the offset is the same as the distance you want to move up, over, or down. See Page 113 if you have difficulty with 90° simple offsets.

Example: If you were running 8" duct and had to offset 27" using 90° long radius elbows, how long would the straight path of duct (center length) be between the elbows? (A long radius elbow has a radius of turn equal to $1\frac{1}{2}$x the width of the duct. In this case, $1\frac{1}{2}$x 8 = 12".)

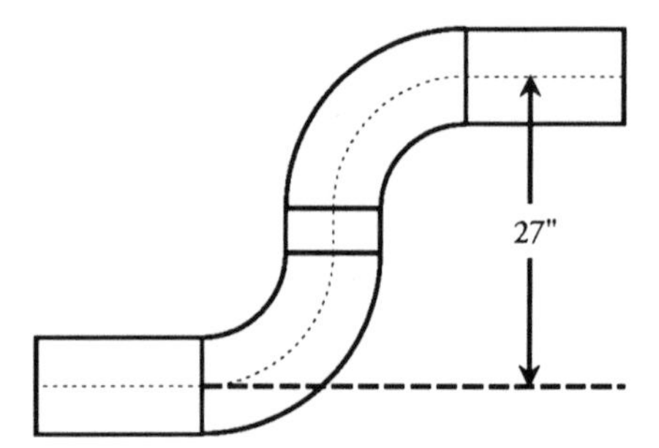

First: Calculate the take out of the elbow.

Take out = tan $\frac{\theta}{2}$ x radius of turn

Take out = tan $\frac{90°}{2}$ x 12"

Take out = tan 45° x 12"

Take out = 1 x 12"

Take out = **12"**

Tan 45° = 1

Since the tangent of a 45° angle is *one*, take outs for all 90° elbows or bends are equal to the radii of the turn. Simple take out calculations is another thing that makes the 90° offsets easy.

Second: Find the center length.

Center length = Hypotenuse - 2 T.O.

Center length = 27" - (2 x 12")

Center length = 27" - 24"

Center length = **3"**

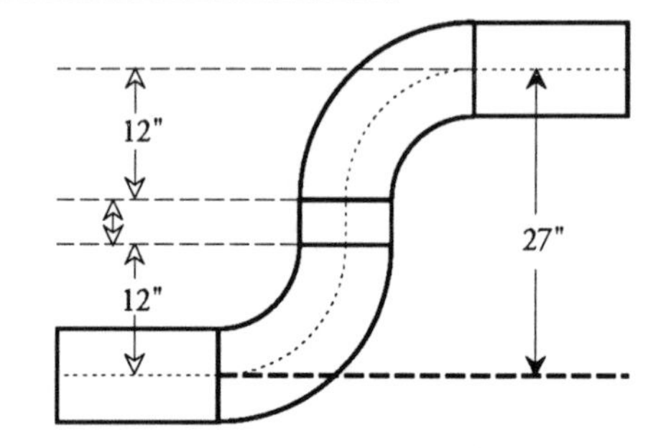

Practice 50: Find the center length for each 90° offset. Round your answers off to four decimal places. Answers-Page 221.

	Offset	Radius		Offset	Radius
(1)	63"	20"	(5)	97"	8"
(2)	18"	6"	(6)	4'6"	12"
(3)	11.4375"	3"	(7)	14'	27"
(4)	21'	18"	(8)	17"	1.5"

30° and 45° Simple Offsets with Arcs.

To determine center length for a 30° or 45° simple offset using arcs:

First: By referring to the prints or your formulations, determine which offset triangle to use—30° or 45°. Then calculate the length of the hypotenuse for the 30° or 45° simple offset.

Second: Calculate the length of the take out for the bend or elbow.

Third: Find the center length of the offset by subtracting two take outs from the length of the hypotenuse.

Example: Blueprints show a 45° simple offset of 84" for a 2' walkway with arcs. The arcs have a radius of 3'. Find the center length.

Remember: List your *knowns* on a thumbnail sketch and note your *needs.*

First: Find the length of the hypotenuse of the walkway using the offset right triangle.

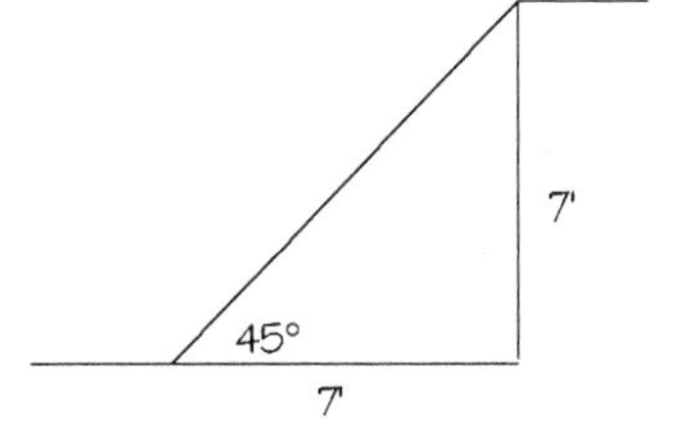

Remember: To find the length of the hypotenuse of a 45° right triangle, multiply a leg by 1.4142.

Hypotenuse = Opposite side x 1.4142

Hypotenuse = 7' x 1.4142

Hypotenuse = 9.8994'

Second: Find the take out for 45° bends.

Take out = $\tan \frac{\theta}{2}$ x radius of turn

Take out = tan 22.5 x 3'   tan 22.5 = .4142

Take out = .4142 x 3'

Take out = 1.2426'

Third: Subtract two take outs from the length of the hypotenuse to find the center length.

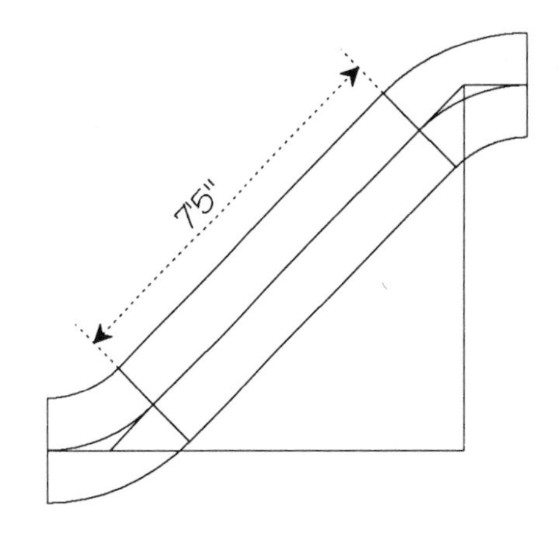

Center length = Hypotenuse - 2 T.O.

Center length = 9.8994'- (2 x 1.2426")

Center length = 9.8994' - 2.4852

Center length = 7.4142' or 7' 5"

Practice 51: Find the center length for these 30° and 45° simple offsets. Round your answers off to four decimal places. Answers-Page 221.

30° Simple Offsets

	Opposite side	Radius
(1)	67.3125"	1'
(2)	17' 7"	19"
(3)	1' 9"	8"
(4)	11.375"	6"

45° Simple Offsets

	Opposite side	Radius
(5)	39' 11"	10'
(6)	42"	6"
(7)	95"	18"
(8)	104.333'	20'

Seeing is Believing: Page 200

notes

Odd Angle Simple Offsets with Arcs

With odd angle simple offsets, the angle of turn is decided by the room you have rather than the angle you want to use. Once you find the angle of turn, follow the regular procedure for finding the center length.

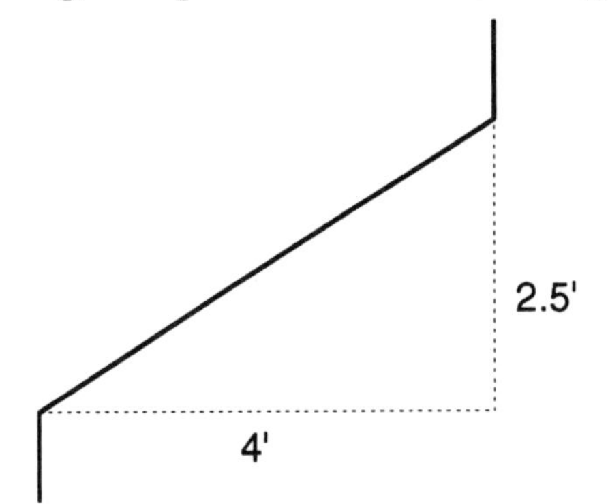

Example: A sheet metal mechanic running a 8" square duct has 2'6" to make an offset of four feet. The specifications call for this line to be installed using long radius elbows.

(Long radius elbows have a radius of turn that is 1.5 times the size of the duct.) 1.5 x 8" = 12"

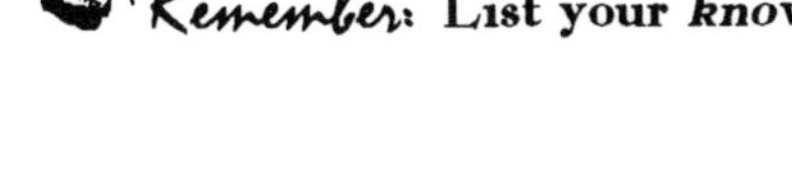
Remember: List your *knowns* on a thumbnail sketch and note your *needs.*

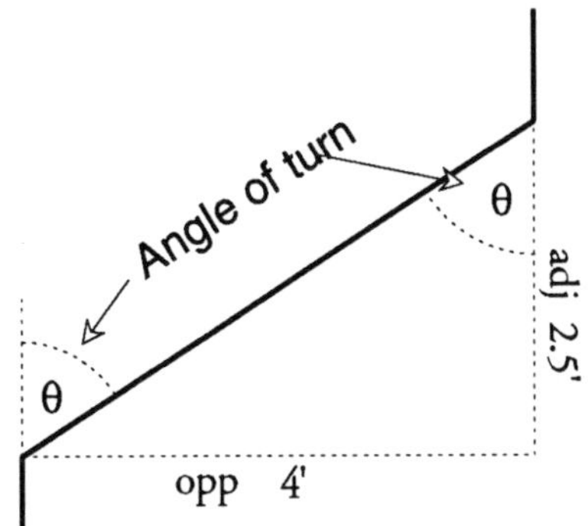

First: Find the angle of turn.

$\text{Tan}\ \theta = \dfrac{\text{opp}}{\text{adj}}$

$\text{Tan}\ \theta = \dfrac{4'}{2.5'}$

$\text{Tan}\ \theta = 1.6$

$\text{Tan}\ 58° = 1.6$

58° is the angle of turn.

Second: Find the length of the hypotenuse.

Hypotenuse = csc θ x opposite side

Hypotenuse = csc 58° x 4'

Hypotenuse = 1.1792 x 4'

Hypotenuse = 4.7168' or **4' 8 $\frac{5}{8}$"**

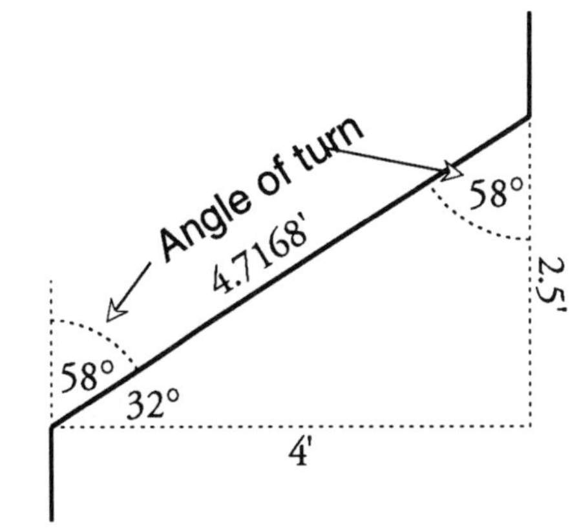

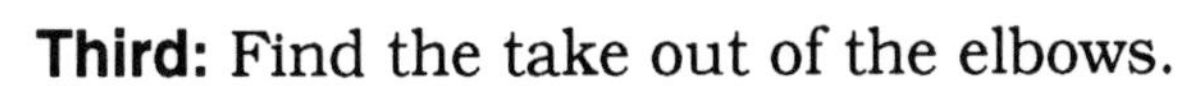

Third: Find the take out of the elbows.

Take out = tan $\frac{\theta}{2}$ x radius of turn

Take out = tan 29° x 12"   tan 29° = .5543

Take out = 12" x .5543

Take out = 6.6516" or **6 $\frac{5}{8}$"**

Fourth: Find the center length.

Center length = Hypotenuse - 2 T.O.

Center length = 4' 8 $\frac{5}{8}$" - (2 x 6 $\frac{5}{8}$")

Center length = 4' 8 $\frac{5}{8}$" - 13 $\frac{1}{4}$" (1' 1$\frac{1}{4}$")

Center length = **3' 7 $\frac{3}{8}$"**

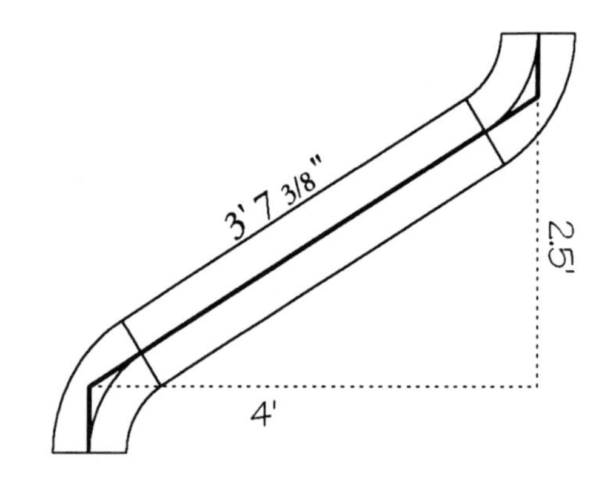

Practice 52: Find the center length for each odd angle simple offset. Round the answers off to four decimal places. Answers-Page 221.

	Radius of turn	Opposite side	Adjacent side		Radius of turn	Opposite side	Adjacent side
(1)	8"	14"	19"	(5)	18"	123.625"	14"
(2)	20"	76"	54"	(6)	3'	3'	4'
(3)	6"	15.5"	23.25"	(7)	150'	56.25'	251'
(4)	30"	34"	89.25"	(8)	24"	19"	49"

Note: On the job, you will find many offsets with fittings that require some manner of attachment to the straight lengths of material. The manner of connection varies among different trades. For example, with butt welded pipe, pipe fitters must allow for a welder's gap of $\frac{3}{32}$" or more. The spacing for each connection must be subtracted from the center length. We are not showing this spacing here because of the wide variety of connectors used in different fields; however, such spacing may need to be considered in your work.

notes

Mitered Turns

A **mitered turn** is made by cutting **miters** (angles) at the ends of the material being used and fastening them together.

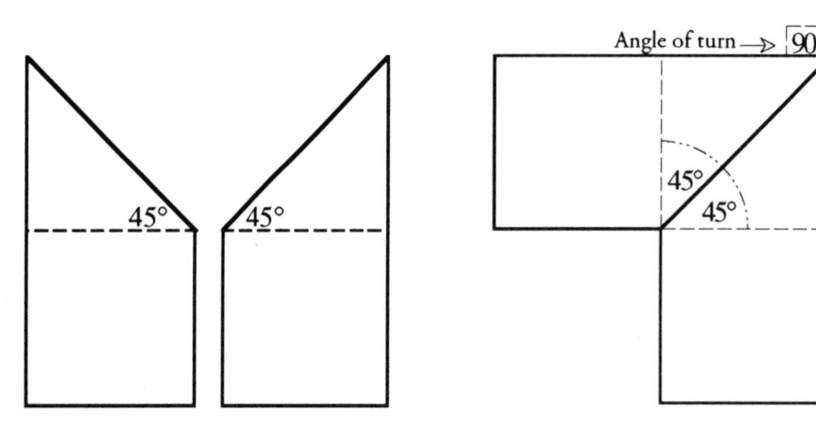

Notice the angle of turn and the angles of the miters in the drawings above. The angle of each miter is equal to the angle of turn divided by two. This is a rule for cutting or marking miters for mitered turns.

To get even miter joints, always divide the angle of turn by 2 to determine the angle of the individual miters.

Example: If a 45° mitered turn is needed, divide 45° by 2 ($\frac{45°}{2} = 22\frac{1}{2}°$).

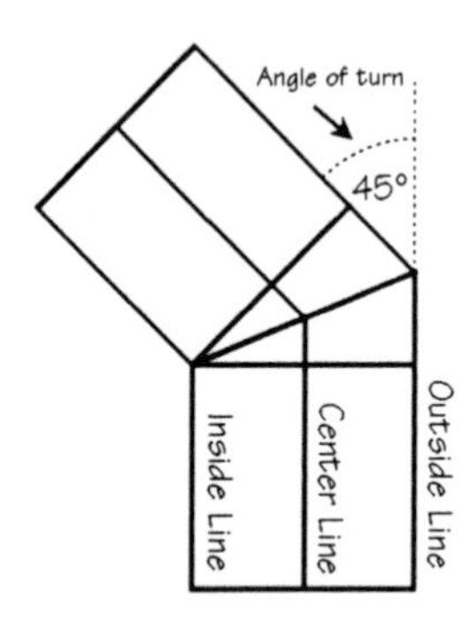

Some trades calculate and mark miters from the center line; therefore, a center line has been added to the drawings. Although many miters are marked with tools like combination squares and protractors, large miters and pipe miters are laid out using the right triangle.

As you can see in the drawing on the right, there are two equal right triangles in each mitered turn. Since these triangles are equal, we can separate them and work with just one. Everything found for one applies to the other.

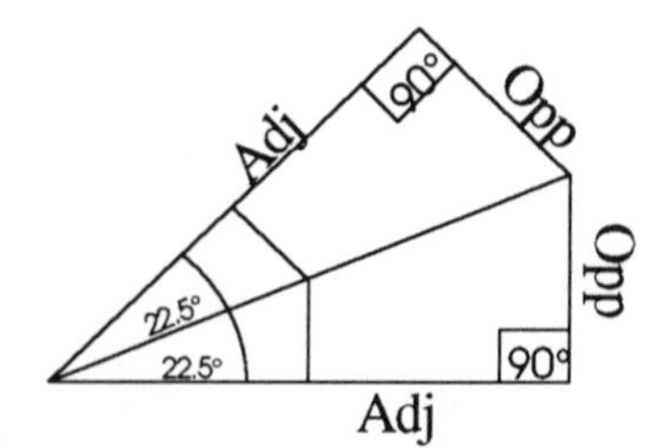

- Notice that the length of the adjacent side of the right triangle is the width of the material.

The *knowns* for this right triangle are

- the reference angle and
- the length of the adjacent side.

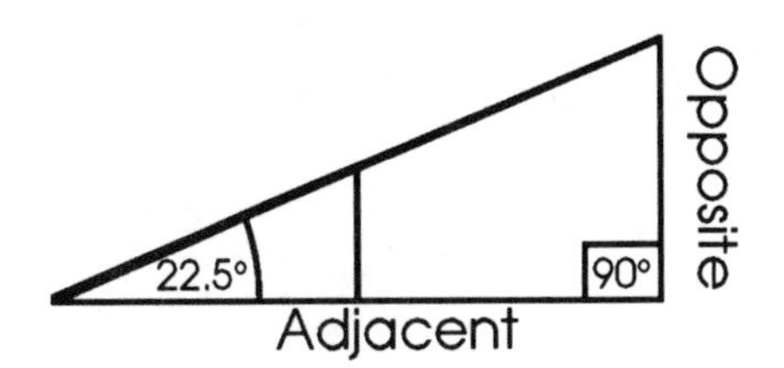

For this example, use 10" as the width of the material (the length of the adjacent side).

NAK Chart				
Need		Angle		Known
Hyp	=	csc θ	x	Opp
Hyp	=	sec θ	x	Adj
Opp	=	tan θ	x	Adj
Opp	=	sin θ	x	Hyp
Adj	=	cos θ	x	Hyp
Adj	=	cot θ	x	Opp

The *need* is

- the length of the opposite side.

By referring to the NAK chart, you find that the formula, **Opposite side = tangent θ x Adjacent side**, uses the *knowns* to find the *need*.

Opposite side = tangent θ x Adjacent side

Opposite side = tan 22.5° x 10"

Opposite side = .414214 x 10"

Opposite side = **4.14214"** = outside length of a miter

The outside length of a miter is also referred to as **Outside Mark Back**.

Outside Mark Back = Tan θ x width

To find the center line length, you can divide the width of the material by two, then use the NAK function again.

Opposite side = tangent θ x Adjacent side

Opposite side = tan 22.5° x 5"

Opposite side = .414214 x 5"

Opposite side = **2.07107"** = center line length of a miter

The center length of a miter is also referred to as **Center Mark Back**.

$$\textbf{Center Mark Back} = \textbf{Tan}\ \theta\ \textbf{x}\ \frac{\textbf{width}}{\textbf{2}}$$

Notice that the length of the outside line is twice the length of the center line. This means that you can also divide the outside length by two and find the center line length.

This is how the triangle looks now.

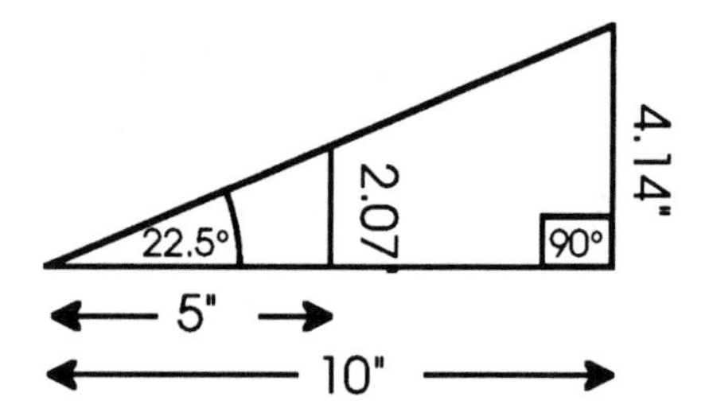

NOTE: The *Outside Mark Back* or the *Center Mark Back* is the distance used to mark a line to cut a miter by.

Miters can be marked using both of the previous measurements (4.14" and 2.07"). Some material is more easily marked by measuring up one side. Other material, especially round objects, is more easily marked by measuring from the center of the object. Below you will find examples of both methods.

Marking Miters Using the Outside Mark Back Measurement

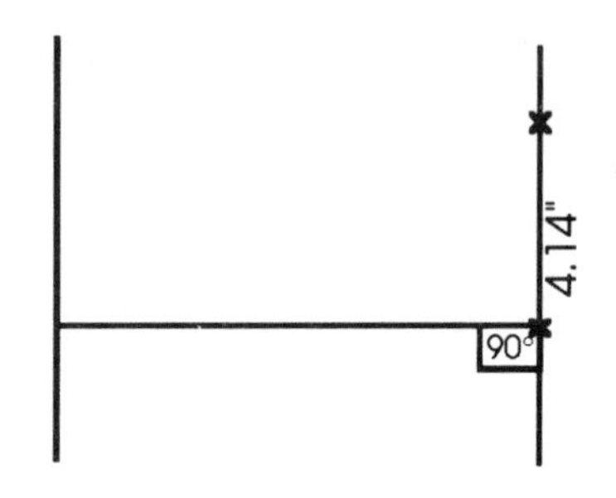

Take a piece of ten inch wide channel iron and mark a square line across it. Measure and mark $4\frac{1}{8}$" above the line on one side of the channel iron.

Mark a new line from the $4\frac{1}{8}$" mark to the end point of the squared line on the opposite side. You will have drawn a right triangle and marked a 22.5° miter.

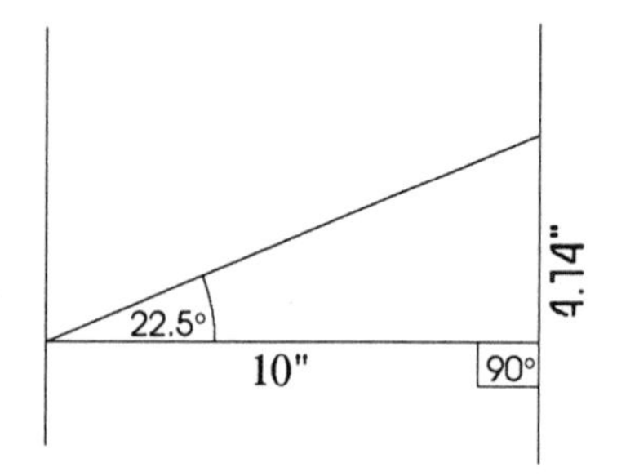

Marking Miters Using Center Mark Back Measurement

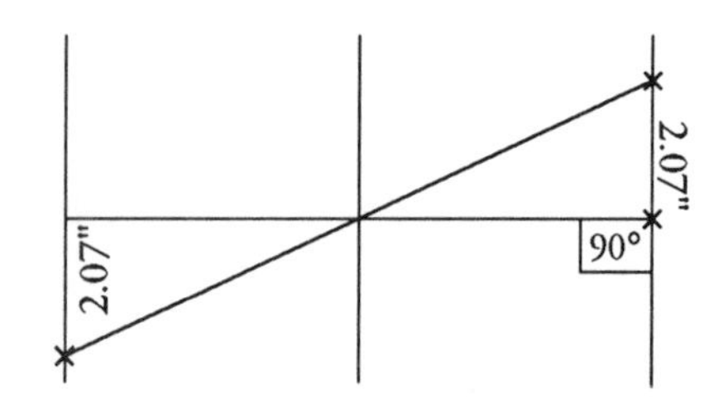

Mark a point on the channel iron where you need the center point to be. Draw a square line across the channel iron through the marked center point. Measure and mark $2\frac{1}{16}$" above the squared line on one side of the channel iron and $2\frac{1}{16}$" below the squared line on the other side. Draw a line between the two marks and you will have marked a 22.5° miter based on a point on the center line.

You will find the center point method useful when working with offsets since the calculations of the offset triangles are based on center lines.

Let's review the process of calculating for miters.

First: Draw a thumbnail sketch and note your *needs* and your *knowns*.

Second: Find the degrees of the miter by dividing the angle of turn by two and calculating the length of the opposite side.

•Opposite side = Tan θ x Adjacent side

Remember: The length of the adjacent side is the width of the material.

Third: Calculate the distance to mark the material for cutting.

Remember: **Outside Mark Back = Tan θ x width**

and

Center Mark Back = Tan θ x $\frac{\text{width}}{2}$

Remember: θ is the angle of the miter

Practice 53: Determine the miter angles and Center Mark Backs and Outside Mark Backs for each of these mitered turns. Round off your answers to four decimal places. Answers-Page 221.

	Angle of turn	Width of material		Angle of turn	Width of material
(1)	60°	8.5"	(5)	45°	26'
(2)	20°	4'	(6)	90°	12.625"
(3)	30°	16"	(7)	57°	7.5625"
(4)	75°	3"	(8)	125°	52"

Offset Turns Using Miters

Let's work a 30°- 60° offset turn with miters. **Remember: With offset turns, unless you are working with 45° turns, there are two *different* angles of turn.**

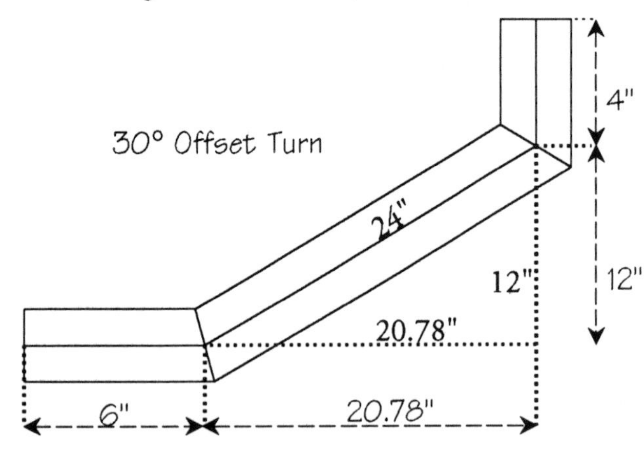

In this turn, the bottom angle of turn is 30° and the top angle of turn is 60°. Together they make a total turn of 90°. Notice that *the center piece of the offset has different lengths for the inside, center, and outside lines.*

The center line is the length of the hypotenuse of the offset triangle.

Except for the upper angle of turn, the sequence for working this offset turn is similar to the one for working a 30° simple offset.

First: Use the offset triangle to determine the length of the hypotenuse and the angles of turn.

Second: Find the angles of the miters by dividing the angles of turn by two.

Third: Calculate Mark Back.
Since the hypotenuse of the triangle we are working from is a center line, use the Center Mark Back formula.
Center Mark Back = Tan θ x $\frac{\text{width}}{2}$

Fourth: Mark your miters.

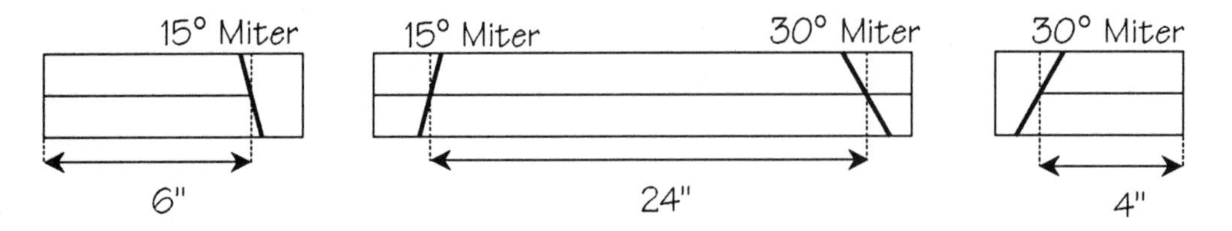

Look at the center piece above. The miters on both ends are marked from the center point. Notice that half the miter extends past the center of the miter. This means that in order to cut these miters, you need a piece of material longer than the length of the hypotenuse of the offset.

Let's say that the width of the material in the earlier sketch is three inches. We already know the angles of turn and the length of the hypotenuse, so let's go on to step two.

•Calculate the mark back for each miter. There are two angles of turn, 60° and 30°.

$$\textbf{Center Mark Back} = \textbf{Tan } \theta \textbf{ x } \frac{\textbf{width}}{\textbf{2}}$$

For a 60° angle of turn (30° miter)

Center Mark Back =	$\tan 30° \text{ x } \frac{\text{width}}{2}$	width = 3"
Center Mark Back =	tan 30° x 1.5"	Tan 30° = 0.57735
Center Mark Back =	0.57735 x 1.5"	
Center Mark Back =	0.8660"	

For a 30° angle of turn (15° miter)

Center Mark Back =	$\tan 15° \text{ x } \frac{\text{width}}{2}$	width = 3"
Center Mark Back =	tan 15° x 1.5"	tan 15° = 0.26795
Center Mark Back =	0.26795 x 1.5"	
Center Mark Back =	0.4019"	

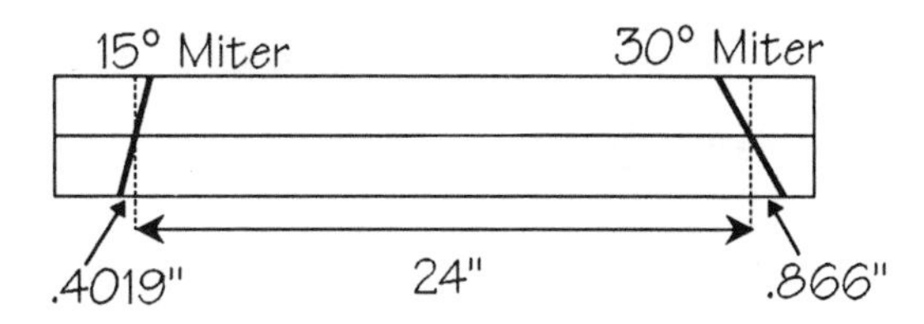

The minimum length of material needed to cut this mitered center piece is 24" + .866" + .402" = 25.268".

Odd Angle Offset Turn

In many situations, you have no control over the angles of turn you must use. Here is an example of an odd angle offset turn.

- In limited space, you must use an offset turn. The material you are using is 6" wide, and you have a 16" adjacent side and a $14\frac{7}{16}$" opposite side.

Find the angle of the miters and the length of the hypotenuse.

First: Draw a thumbnail sketch.

Second: Find the angles of turn.

$$\text{Tangent } \theta = \frac{\text{opposite side}}{\text{adjacent side}}$$

$$\text{Tangent } \theta = \frac{14.4375''}{16''}$$

$$\text{Tangent } \theta = .90235$$

$$\text{Tangent } \mathbf{42}^\circ = .90235$$

The lower angle of turn is 42°.

The upper angle of turn is 90°- 42° = **48°**.

Third: Use the Pythagorean theorem to find the length of the hypotenuse.

$$c = \sqrt{a^2 + b^2}$$

$$c = \sqrt{16^2 + 14.4375^2}$$

$$c = \sqrt{256 + 208.441}$$

$$c = \sqrt{464.441}$$

$$c = \mathbf{21.55''}$$

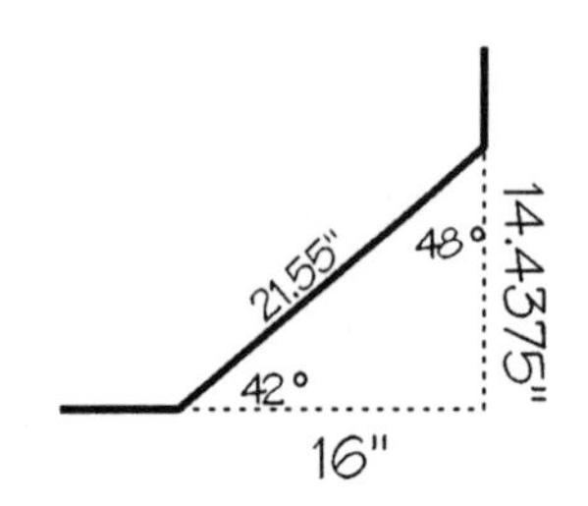

Fourth: Calculate the Center Mark Back. The angles of turn are 42° and 48°.

For a 42° angle of turn (21° miter)

Center Mark Back = $\tan 21^\circ \times \frac{\text{width}}{2}$ Width = 6"

Center Mark Back = tan 21° x 3" Tan 21° = 0.3839

Center Mark Back = 0.3839 x 3"

Center Mark Back = **1.1517"**

For 48° angle of turn (24° miter)

Center Mark Back = $\text{Tan } 24^\circ \times \frac{\text{width}}{2}$ Width = 6"

Center Mark Back = Tan 24° x 3" Tan 24° = 0.4452

Center Mark Back = 0.4452 x 3"

Center Mark Back = **1.3356"**

The minimum length needed for the center piece is 24.04".

21.55" + 1.1517" + 1.3356" = 24.0373"

Practice 54: Find the angles of miter and the minimum length of the center piece for each offset turn below. Round the angles off to the nearest tenth of a degree and the minimum lengths to four decimal places. Answers-Page 221.

	Opposite side	Adjacent side	Lower Angle	Width of Material		Opposite side	Adjacent side	Lower Angle	Width of Material
(1)	84"	46"		14"	(5)	12"	16"		10"
(2)	5'		45°	4'	(6)	62.5"		45°	19"
(3)	39"	39"		9"	(7)	41'	20.5'		12'
(4)	27.5'		30°	6'	(8)	54"		72°	9"

notes

Mitered Offsets

A mitered simple offset is a bit easier to calculate than a simple offset with arcs since there are no take outs involved. *With mitered offsets, all center piece lines are the same length as the hypotenuse of the offset triangle, even though both ends are mitered.*

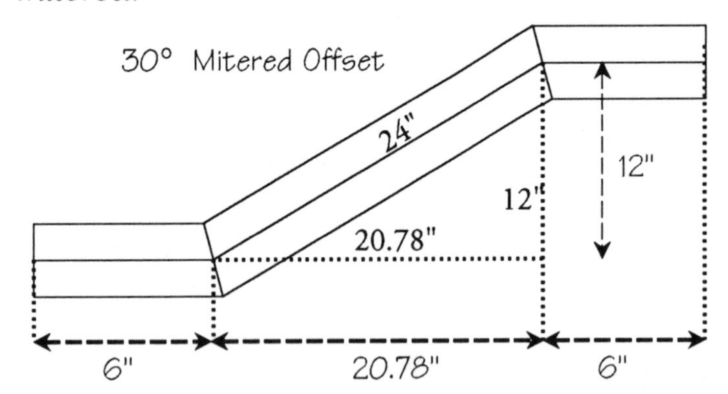

This is a 30° simple offset, so both angles of turn are 30°, and all of the miters are 15° cuts. The offset triangle is a 30°- 60° right triangle.

To determine the measurements needed to cut a mitered offset:

First: Draw a thumbnail sketch. Determine the angles of the miters. Note the *knowns* and the *needs.*

Second: Determine the hypotenuse length for the offset triangle. The center piece can be marked by this measurement.

Third: Determine the Center Mark Back.
$$\textbf{Center Mark back} = \textbf{Tan}\ \theta\ \textbf{x}\ \frac{\textbf{width}}{\textbf{2}}$$

Fourth: Mark and cut the miters.

Example: Find the angle of the miters, the length of the hypotenuse, the Center Mark Back, and the length of the hypotenuse of a 45° simple offset using a material that is 9 inches wide. The distance to be offset is 8".

First: If the simple offset is a 45° simple offset, then the miters are half of that—**22.5°**.

Second: Since the distance you are offsetting is 9", the length of the hypotenuse is 1.4142 x 9 or **12.73"**.

Third: Using the working formula for the Center Mark Back with a width of 8" produces and answer of 1.657":

$$\textbf{Center Mark back} = \textbf{Tan } \theta \textbf{ x } \frac{\textbf{width}}{\textbf{2}}$$

$$\text{Center Mark Back} = \text{Tan } 22.5 \text{ x } \frac{8''}{2} \qquad \text{Tan } 22.5^\circ = .4142$$

$$\text{Center Mark Back} = .4142 \text{ x } 4''$$

$$\text{Center Mark Back} = 1.657''$$

Practice 55: Work each simple offset below using mitered joints. Determine the miter angles, Center Mark Back measurements and the length of the hypotenuses. Round the angles off to two decimal places and the lengths to four decimal places. Answers-Page 221.

	Angle	Opposite side	Width		Angle	Opposite side	Width
(1)	30°	23"	4"	(5)	22.5°	68"	5"
(2)	45°	14"	8"	(6)	30°	9'	5'
(3)	90°	8'	4'	(7)	45°	18.5"	4.75"
(4)	65°	34"	12"	(8)	90°	10'	6"

Notes On Miters

When a miter is cut, both pieces of material are mitered. This is a 15° miter cut.

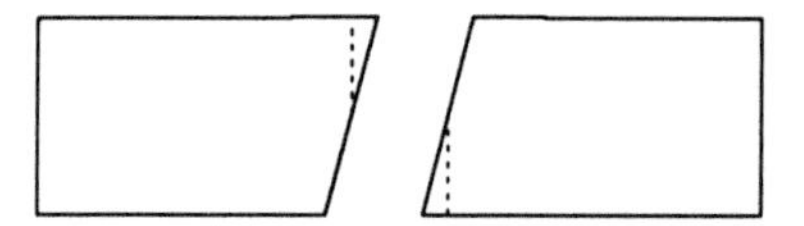

With some material, you will be able to turn one of the pieces over and join it to the other piece to make a mitered joint.

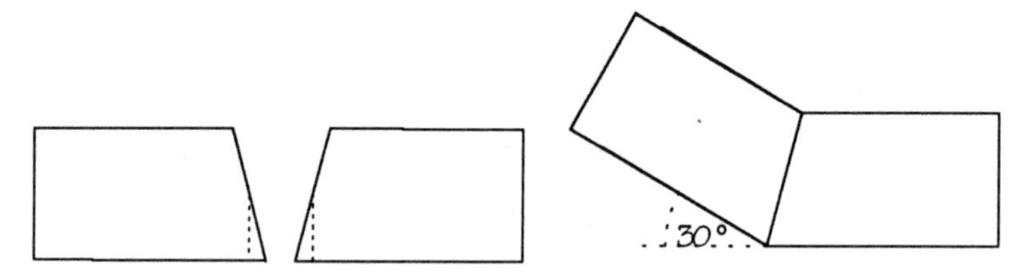

With some material, you might not be able to use the mitered pieces when they are turned over. This depends on whether the piece of material when it is turned over has the same shape as the other piece. Imagine the shapes below to be end pieces of material.

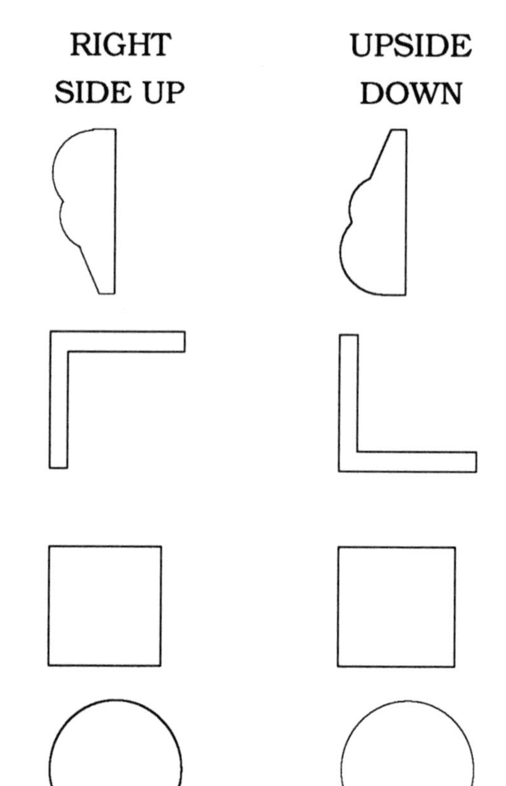

The first two examples have different shapes when turned upside down. Therefore, you would have to cut two separate miters to make a mitered joint.

For the last two examples, either way they are turned, both pieces are the same shape. You could cut a miter, turn one piece over, and join the two pieces to make a mitered joint.

Of course, this method is applicable only when both sides of the material you are working with are usable.

notes

Area

Area is the measurement of the surface of an object. Area is generally used in the trades when the amount of surface must be known for further calculations, for instance to order material. We buy paint based on the area we need to cover. We buy tile and carpet the same way. Roofers determine how much material to purchase for a job by measuring area and surveyors provide land owners with measurements of the area of their property.

Square units of measurement (such as square inches, square feet, and square yards) are used to describe the amount of surface area. The area of the square below is one square inch. 1 inch x 1 inch = 1 square inch.

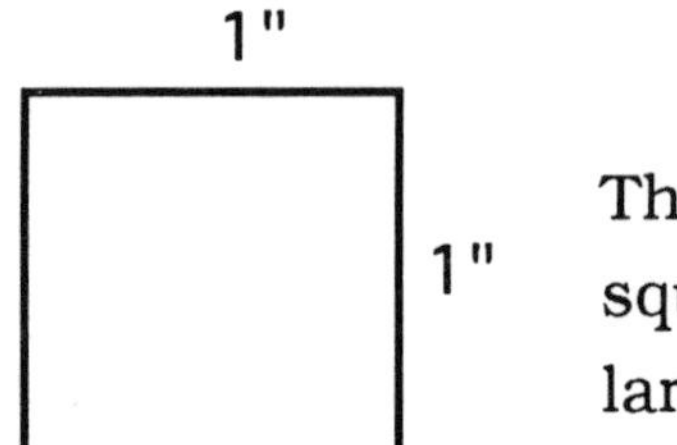

The square inch is the base for small area measurements. The square foot, the square yard, and the square mile are used for larger area measurements.

Rectangles

This page is a rectangle. It is 11" long by $8\frac{1}{2}$" wide.

To find the area of this page, multiply the length by the width.

Area of a rectangle = length x width **A = l x w**

A = 11" x 8.5"

A = 93.5 square inches or in^2

The dimensions of this page can also be expressed in feet. Therefore, the area of this page can be expressed as square feet.

First: Convert the dimensions to a common unit of measurement.

$$\boxed{11"} = \frac{11"}{12" \text{ per foot}} = 0.9167' \quad \boxed{8.5"} = \frac{8.5"}{12" \text{ per foot}} = 0.7083'$$

Second: Calculate the area using the formula A = l x w.

$$A = 0.9167' \times 0.7083'$$
$$A = 0.6493 \text{ square ft. or ft}^2$$

By converting the same dimensions to yards, you can also express the area of this page in square yards.

$$\boxed{11"} = \frac{11"}{36" \text{ per yard}} = 0.3056 \text{ yards} \quad \boxed{8.5"} = \frac{8.5"}{36" \text{ per yard}} = 0.2361 \text{ yards}$$

$$A = l \times w$$
$$A = 0.3056 \text{ yds} \times 0.2361 \text{ yds}$$
$$A = 0.07216 \text{ square yards or yds}^2$$

There are a few things that you should note here.

- **One**: The size of the page has not changed, only the description of the surface area.
- **Two**: Remember from the section on squaring numbers that a number multiplied by itself is the square of that number. The same holds true for units of measurement.

$$\text{Inches} \times \text{inches} = \text{inches}^2$$
$$\text{Feet} \times \text{feet} = \text{feet}^2$$
$$\text{Yards} \times \text{yards} = \text{yards}^2$$

- **Three**: You cannot mix units of measurement and find the correct area. You *cannot* multiply *feet by inches* or *feet by yards* or *inches by yards*. If you are working with mixed dimensions, you must *always* convert them to a common unit of measurement.

Practice 56: Find the area for each rectangles. **A = l x w**. Round your answers off to three decimal places. Answers-Page 221.

	length	width		length	width		length	width
(1)	7"	3"	(3)	6"	2'	(5)	55' 9"	47' 3"
(2)	25'	3'	(4)	7'	6' 2"	(6)	100'	100'

Squares

Since a square is a rectangle, the same formula used to find the area of a rectangle **($A = l \times w$)** can be used for the square; however, since all sides of a square are equal, you can also just square the length of one side.

Area of a square = one side squared
$A = l \times w$ or $A = s^2$

Example: Find the area of a $4\frac{1}{2}$" square using both formulas.

$A = l \times w$	$A = s^2$
A = 4.5" x 4.5"	A = (4.5")^2
A = 20.25 in^2	A = 20.25 in^2

Practice 57: Find the area for each square. Round the answers off to two decimal places. Answers-Page 221.

	Side length		Side length
(1)	3'	(5)	37'
(2)	$22\frac{9}{16}$"	(6)	21' 3"
(3)	19"	(7)	2' 2"
(4)	110'	(8)	6' 11"

Triangles

The formula for determining the area of a triangle is $\frac{1}{2}$ bh. The abbreviations b and h represent base and height.

The base of a triangle is the side it sits on.

The height is the length of the triangle from its base to its highest point.

Area of a triangle = $\frac{1}{2}$ base x height
$A = \frac{1}{2}bh$

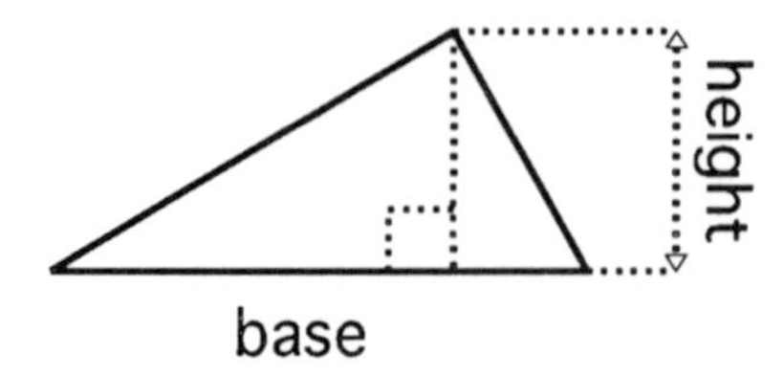

The highest point of a triangle is the vertex of the angle opposite the base. The height of a triangle is the length of a perpendicular line which has been dropped from the vertex of the opposite angle to the base.

Note: Since you decide which side of the triangle is the base, use the side that makes finding the area easiest.

Right Triangles

To determine the area of a right triangle, you should use one of its legs—not the hypotenuse—as the base. If one leg is the base, then the other leg is the height since *it is* a perpendicular line which drops from the vertex of the opposite angle to the base.

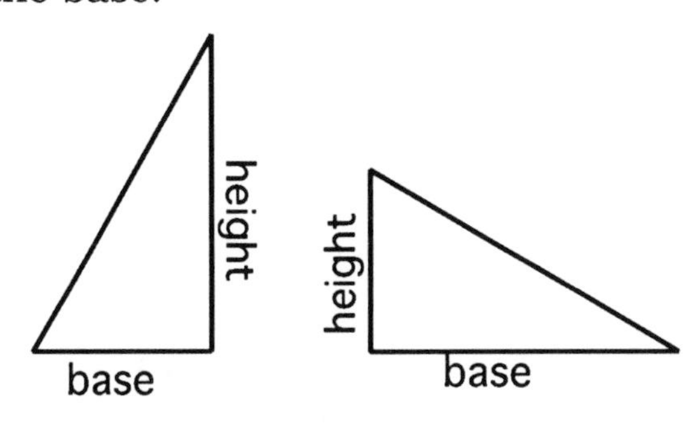

Earlier in the book, you learned that two equal right triangles are created when a diagonal line is drawn through a rectangle. If two triangles make up the total area of the rectangle, then one of those right triangles must cover half that area.

Look at this rectangle divided by a diagonal line.

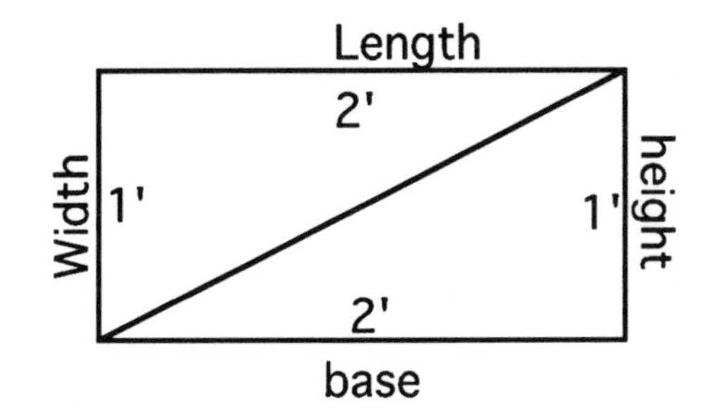

The area of the rectangle is length x width or $1' \times 2' = 2$ ft^2.

The area of each right triangle is $\frac{1}{2}$ bh or $\frac{bh}{2}$. $\frac{2' \times 1'}{2} = 1$ ft^2.

For example: Find the area of each right triangle below.

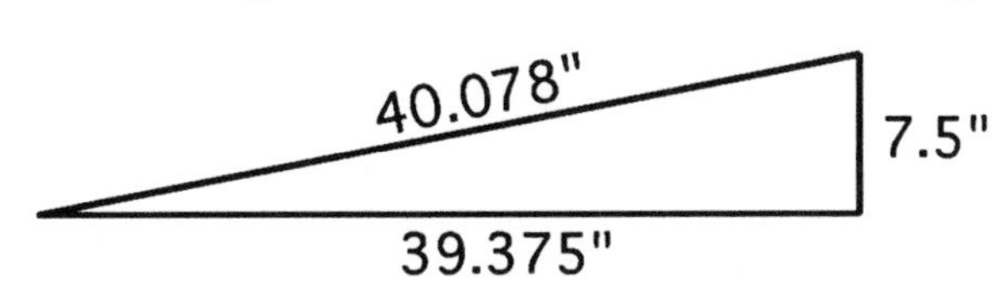

$A = \frac{1}{2}bh$

A = 0.5 x 39.375" x 7.5"

A = 147.656 square inches

Practice 58 Find the area for each right triangle. Round off your answers to two decimal places. Answers-Page 221.

	base	height		base	height
(1)	9'	12'	(5)	$23\frac{1}{2}$"	20"
(2)	15'	20'	(6)	103' 3"	76' 7"
(3)	21"	33"	(7)	$17\frac{15}{16}$"	$23\frac{15}{16}$"
(4)	7' 9"	3' 4"	(8)	16.73"	27.32"

Isosceles Triangles

Remember: **An isosceles triangle has two equal sides.**

When determining the area of an isosceles triangle, *use the unequal side as the base.* The height of the isosceles triangle is the length of a bisecting line which drops from the vertex of the angle opposite the base to the base. That bisecting line also creates two right triangles.

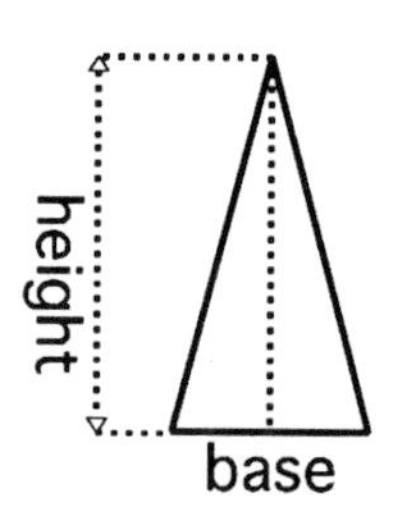

Let's say our sample isosceles triangle has a central angle of 30° and two equal sides of 20". That means that the *knowns* for each right triangle are:
The length of the hypotenuse, 20" and the reference angle of 15°.
In order to use the triangle formula for area, $A = \frac{1}{2}bh$, you *need* to know the length of the adjacent side (height) and the length of the opposite side ($\frac{1}{2}$ the base).

Use the NAK chart to find the correct function to use.

Hypotenuse	=	csc θ	x	Opposite
Hypotenuse	=	sec θ	x	Adjacent
Opposite	=	tan θ	x	Adjacent
Opposite	**=**	**sin θ**	**x**	**Hypotenuse**
Adjacent	**=**	**cos θ**	**x**	**Hypotenuse**
Adjacent	=	cot θ	x	Opposite

First: Find the length of the adjacent side.

Adj. = cos 15° x Hypo.

Adj. = 0.9659 x 20"

Adj. = 19.318"

Height = **19.318"**

Second: Find the length of the opposite side.

Opp. = sin 15° x Hypo.

Opp. = 0.2588 x 20"

Opp. = 5.176"

Base = 2 x 5.176" = **10.352"**

Notice that the length of the opposite side is multiplied by *two* to find the length of the base.

Third: Calculate the area.

Area = $\frac{1}{2}$ base x height

Area = 0.5 x 10.352" x 19.318"

Area = **100 square inches**

Practice 59: Find the area for each isosceles triangle. Round your answers to two decimal places. Answers-Page 221.

	Odd angle	equal side length		Odd angle	equal side length
(1)	60°	12"	(4)	15°	2.5'
(2)	90°	6'	(5)	35.5°	$3\frac{1}{8}$"
(3)	45°	3'	(6)	130°	72'

Circle

The formula for the area of a circle is πr^2. Somewhere in school, in order to help me remember this formula, a teacher said, "Pi are square, cakes are round." It stuck with me and maybe it will with you.

Area of a circle = πr^2

Example: Find the area of a 12' diameter circle.

$A = \pi r^2$

$A = 3.1416 \times (6')^2$

$A = 3.1416 \times 36$ sq. ft.

$A = 113.0976$ sq. ft.

• Note: Remember that the formula calls for the **radius²**, not the diameter². Squaring the diameter is probably the most common mistake people make when using this formula. Squaring the diameter instead of the radius results in an answer four times larger than the correct answer. The radius in the above example is 6. Look at the difference if you mistakenly square the diameter measurement: Not $6^2 =$ **36** , but $12^2 =$**144.**

Practice 60: Find the area for each of the following circles. Round the answers off to two decimal places. Answers-Page 221.

	Radius		Radius
(1)	20'	(5)	13' 6"
(2)	100"	(6)	3 yds, 2' 1"
(3)	15.5 yds	(7)	.5 miles
(4)	15.5 "	(8)	12' 10"

Side of a Cylinder

Painters on industrial jobs are always painting tanks and they buy paint based on the amount of surface they need to cover. The formula for determining the area of a circle will work for the top of the tank, but the sides require a different formula. Let's take a look.

Example: Find the area of the side of a circular tank (cylinder) 10 feet in diameter and 20 feet high.

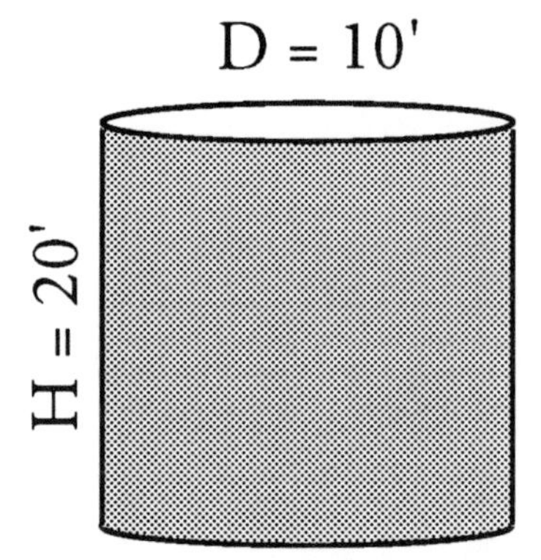

If you could unroll this tank, you would see it as a rectangle.

The length of the rectangle is πd (circumference of the top of the tank) or 3.1416 x 10' or 31.416'.

The height of the tank is the width of the rectangle.

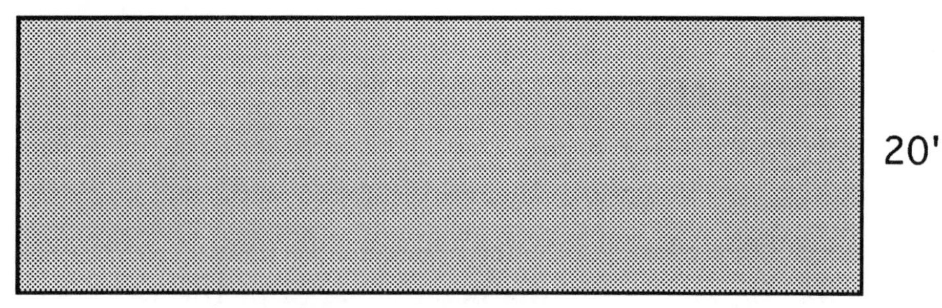

Finding the area is now a matter of calculating the area of a rectangle.

Area of rectangle = length x width

Area = 31.416' x 20'

Notice that you can combine the two formulas and say that the area of the side of a cylinder equals πdh.

Practice 61: Find the surface area for each cylinder below. Round your answers off to two decimal places. Answers-Page 222.

	diameter	circumference	height		diameter	circumference	height
(1)		100'	25'	(5)		365.5'	55"
(2)		10"	10"	(6)		122.5'	20'
(3)	10'		10'	(7)	2' 9"		5'
(4)	4.5"		4.5"	(8)	$2\frac{5}{16}$"		1.25'

notes

XIII

Volume

Volume is defined as the amount of space occupied in three dimensions.

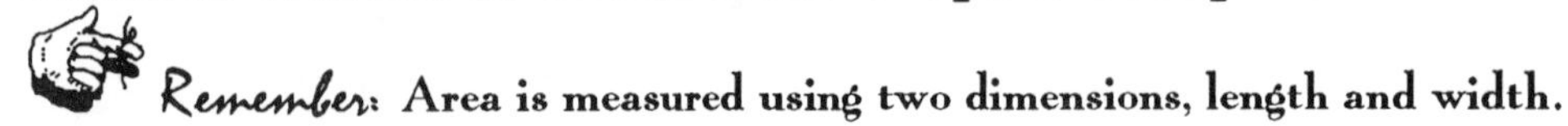

Volume is measured using three dimensions, length, width and height (or depth or thickness). The units used in measuring are the cubic inch (in^3), the cubic foot (ft^3), and the cubic yard (yd^3).

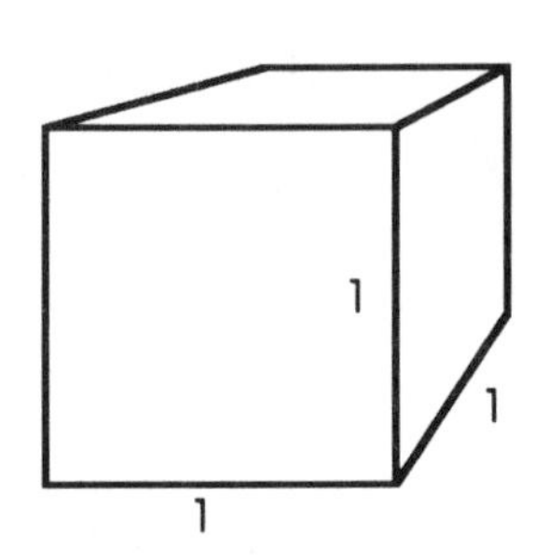

Cubic measurements are based on the cube.

A cube with 1" sides will measure one cubic inch.

1" x 1" x 1" = 1 in^3 = One cubic inch

A cube with 1' sides will measure one cubic foot.

1' x 1' x 1' = 1 ft^3 = One cubic foot

A cube with 1 yd sides will measure one cubic yard.

1 yd x 1 yd x 1 yd = 1 yd^3 = One cubic yard

Rectangular Prism

This page is $8\frac{1}{2}$" x 11" x $\frac{1}{456}$". It is a rectangular prism, which means that it is a rectangle with thickness. (A fancy word for a box)

Volume of rectangular prism = length x width x height **V = l x w x h.**

H stands for height, depth or thickness.

Volume of this page = 11" x 8.5" x 0.002193"
0.205 cubic inches

- We used two dimensions, length and width, to find the area of the surface of this page, length and width. A = **l x w**. Notice that the formula for volume uses the same dimensions as the formula for area, plus one more—h.

Volume = l x w **x h**.

The area of a surface can be multiplied by the depth or height to find the volume.

Area x Height = Volume

93.5 sq in(see page 147) x 0.002193" = 0.205 cubic inches
(Multiplying square inches by inches produces cubic inches.)

- We could express the dimensions of this page in feet in order to find its volume in cubic feet, and in yards in order to find its volume in cubic yards.
- As usual, in your calculation, use common units of measurement for all dimensions.

Calculating the Correct Amount of Concrete to Order

The formula for determining the volume of a rectangular prism as noted above, is **length x width x height**. When I think of the volume of a rectangular prism, I almost always think of concrete, because most concrete is poured into rectangular forms.

Concrete is sold and ordered by the cubic yard. Few rulers show yards, so generally the dimensions of forms are measured in feet and inches, then converted to cubic yards.

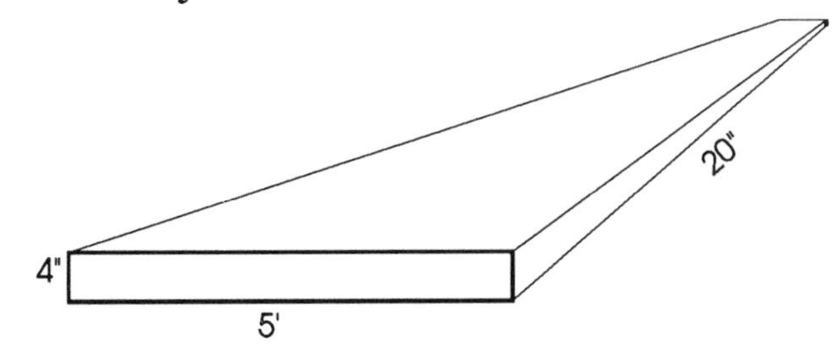

This sidewalk is 5 feet wide, 20 feet long, and 4 inches deep. How many cubic yards of concrete will it take to pour this side walk?

Convert the inches to feet to give you a common unit of measurement:

$$\frac{4"}{12" \text{ per foot}} = 0.333'$$

Volume = length x width x height

Volume = $5' \times 20' \times 0.333' = 33.3\ ft^3$

(There are 27 ft^3 in a cubic yard) ($3' \times 3' \times 3' = 27\ ft^3$)

$$\frac{33.3\ ft^3}{27\ ft^3 \text{ per } yd^3} = \mathbf{1.23\ yds^3}$$

Practice 62: Find the volume of each rectangular prism below. Answers-Page 222.

Remember: **Volume of a rectangular prism = Length x width x height**

	length	width	height		length	width	height
(1)	36"	22"	3"	(5)	55"	.5'	2.5'
(2)	333'	20'	3'	(6)	.005"	.1275"	1.25"
(3)	22 yds	22 yds	22 yds	(7)	$3\frac{1}{2}$	9.5'	2.775"
(4)	20 miles	10 miles	2 miles	(8)	1 yd	1"	1'

Cubes

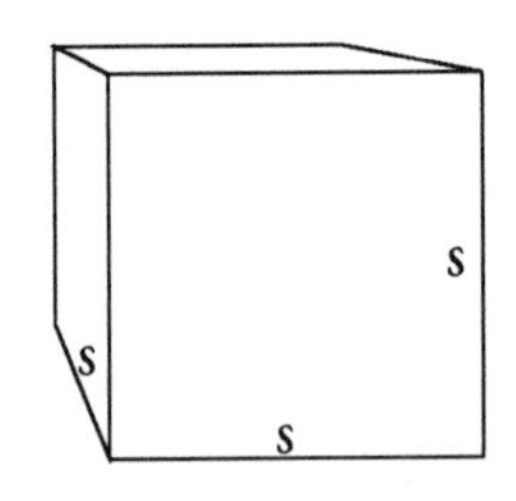

The cube is a special kind of rectangular prism because its length, width, and height are equal. To find the volume of a cube, you can use the formula for rectangular volume, but it is easier simply to cube one of the dimensions.

Volume of a cube = s^3

Remember: **To cube a number, multiply it by itself twice.** $s \times s \times s$.

Example: Dice cubes measure $\frac{1}{2}$" per side. What is the volume of one die*?

Volume of cube $= s^3$

Volume = $(\frac{1}{2}")^3$

Volume = 0.5" x 0.5" x 0.5"

Volume = 0.125 in^3

Practice 63: Find the volume of each cube below. Answers-Page 222.

	side length		side length		side length		side length
(1)	1"	(3)	22 miles	(5)	$1\frac{1}{2}'$	(7)	16.5"
(2)	.05'	(4)	10.5 yds	(6)	99"	(8)	122.375'

* Die is the singular of dice. You have one die, but two or more dice.

Triangular Prism

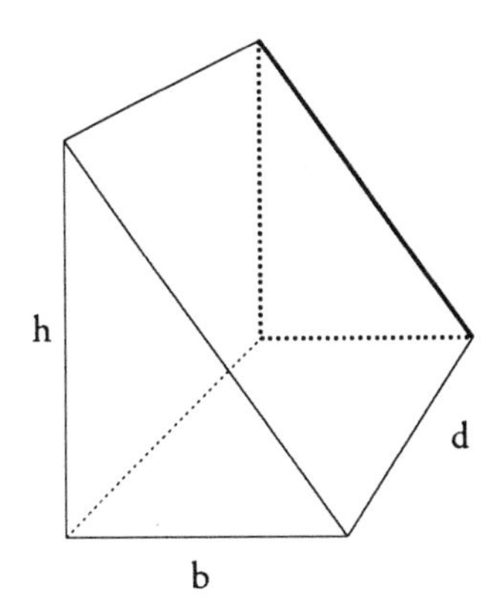

We will work triangular volume for only the right triangular prism. Probably the easiest way to identify the volume of a right triangular prism is to express it as the area *of a right triangle* x *depth.* You can express the formula as:

$$\text{Triangular volume} = \frac{1}{2} b \times h \times d$$

d represents depth or thickness. The triangle formula for area already includes **h,** which represents the height of the triangle.

Example: Find the volume of this right triangular solid. The legs of the right triangle are 6" and 12" long and the prism stands 11" tall.

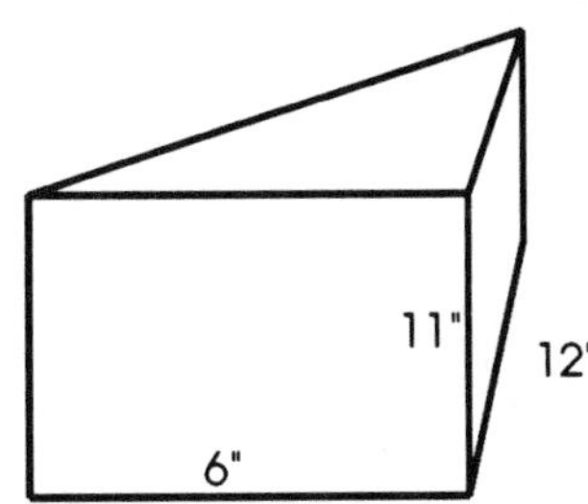

First: Find the area of the triangle.

Area of right triangle $= \frac{1}{2} b \times h$

Area of right triangle = 0.5 x 6" x 12"

Area of right triangle = 36 Sq. in.

Second: Multiply the area measurement by the depth of the triangular prism.

Volume of right triangular solid = area of right triangle x depth

Volume of right triangular solid = 36 sq. in. x 11"

Volume of right triangular solid = **396 in^3**

Practice 64: Calculate the volume for each of the following right triangular solids. Answers-Page 222.

	Leg length	Leg length	Height		Leg length	Leg length	Height
(1)	3'	4'	5'	(4)	2'	1' 6"	3'
(2)	7"	12"	15"	(5)	12"	2'	3'
(3)	9'	7'	22'	(6)	2.5'	16"	$\frac{1}{2}$ yd

Cylinders

Volume of a cylinder = $\pi r^2 h$

Notice this formula is the area of the circle(πr^2) x height(h).

Example: Find the volume of this cylinder in cubic feet.

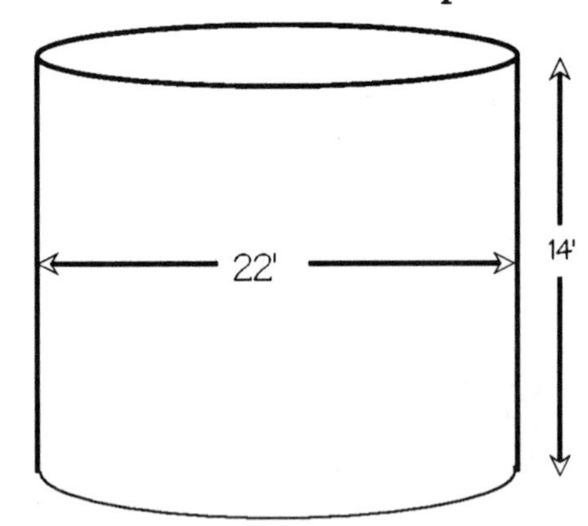

Area of circle = πr^2

Area of circle = 3.1416 x $11'^2$

Area of circle = 380.1336 sq. ft.

Volume of cylinder = area x h

Volume = 380.1336 ft^2 x14 ft

Volume = 5321.8704 ft^3

Practice 65: Find the volume for each cylinder below. Answers-Page 222.

	radius	height		radius	height
(1)	1"	2"	(5)	$25\frac{3}{4}$"	$50\frac{1}{2}$"
(2)	2"	1"	(6)	100.5"	62.5 "
(3)	10'	.5"	(7)	12"	30'
(4)	66 yds	66 yds	(8)	$16\frac{15}{16}$"	$42\frac{7}{8}$"

notes

XIV

Equal Sectors and Regular Polygons

Equal Sectors

In this section, we will use the concept of equal sectors to help you understand more about the properties the circle. If you divide a circle into equal sectors, it looks like a pie. Pies in mathematics are not new to us. We've seen fractions described with pies and ratios and percentages illustrated with pie charts.

Dividing the circle into equal sectors is not difficult. You need to consider only two factors of the circle—the central angle and the circumference. To create equal sectors, divide the number of sectors into both of these factors.

$$\frac{360°}{\text{number of sectors}} \text{ and } \frac{\text{circumference}}{\text{number of sectors}}$$

This gives you the degrees of the central angle of each sector and the arc length of each sector.

Example: Divide a circle with a radius of 10" into eight parts or sectors.

First: Find the angle of the sector by dividing 360° by the number of sectors.

$$\frac{360°}{8} = \mathbf{45°}$$

Second: Find the arc length of each sector by dividing the circumference by the number of sectors.

$$\frac{2\pi r}{8} = \frac{2 \times 10" \times 3.1416}{8} = \frac{62.832"}{8} = 7.854" \text{ or } 7\frac{\mathbf{27}}{\mathbf{32}}"^*$$

* Notice that thirty-seconds are being used now. When marking around circles, a higher degree of accuracy is needed.

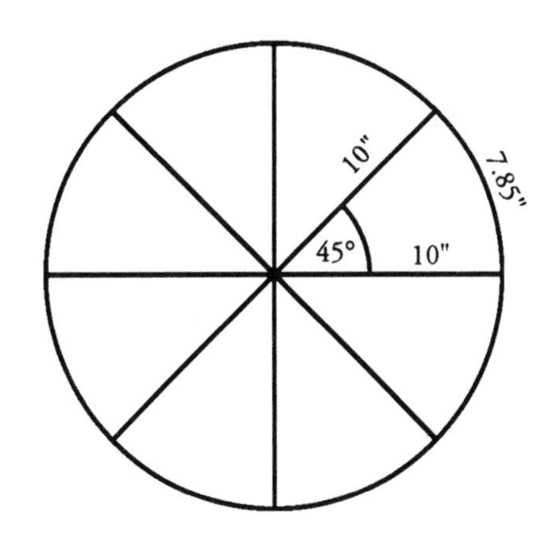

As you can see, there are eight equal sectors in this circle. Each sector has the same dimensions—10" radii, 45° central angles, and $7\frac{27}{32}$" arc lengths.

If this were a 20" diameter pipe, you could divide it into eight equal parts by marking around the outside of the pipe every $7\frac{27}{32}$". In pipe fabrication and tank building, this is an important procedure. It allows you to leave the side of a tank or a pipe at a certain angle by measuring and marking the outside of the tank or pipe.

Practice 66: Find the central angle and arc length of each of the following equal sectors. Answers-Page 222.

	radius	circumference	# sectors		radius	circumference	# sectors
(1)	12"		6	(4)	$158\frac{1}{2}$ miles		8
(2)		22"	10	(5)		$\frac{1}{2}$'	4
(3)		112'	3	(6)	12.555'		12

Marking Equal Sectors

Dividing pipe or tanks into equal sectors is a fairly straightforward procedure since you can measure around the cylinder. The approach to marking equal sectors is a little different when you cannot do such measuring. Carpenters building forms for tank bases or placing anchor bolts for vessels and pipe fitters marking bolt holes for flange fabrication have this problem. They are given the diameter of a circle and must accurately mark equal sectors around the circle. It is close to impossible for them to keep a ruler flexed evenly to measure arcs on a flat surface, so they use chords, not arcs, to mark their equal sectors.

Just as every sector has a precise arc length, every sector has a precise chord length. Let's use the same example introduced above and find the chord length for a circle (r = 10") that has been divided into eight sectors.

From the section on chords, you know that chord lengths are found by using the formula,

$$\textit{Chord} = \sin\frac{\theta}{2} \times \textit{radius} \times 2$$

or

$$\textit{Chord} = \sin\frac{\theta}{2} \times \textit{diameter}$$

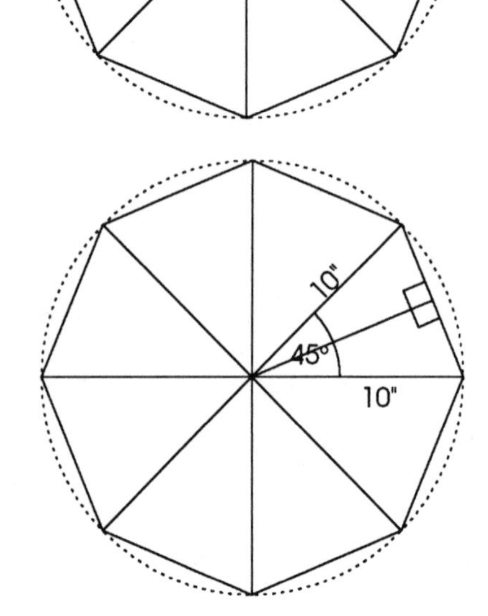

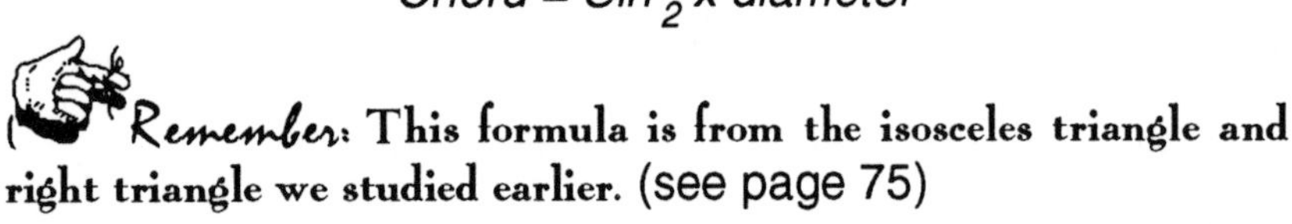

Remember: This formula is from the isosceles triangle and right triangle we studied earlier. (see page 75)

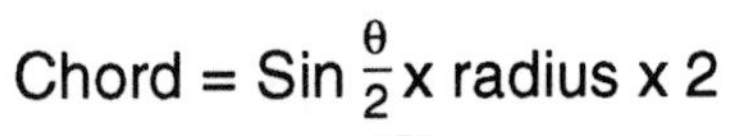

Chord = $\sin\frac{\theta}{2}$ x radius x 2

Chord = $\sin\frac{45°}{2}$ x 10" x 2

Chord = Sin 22.5° x 20"

Chord = 0.38268 x 20"

Chord = 7.6536" or **$7\frac{21}{32}$"**

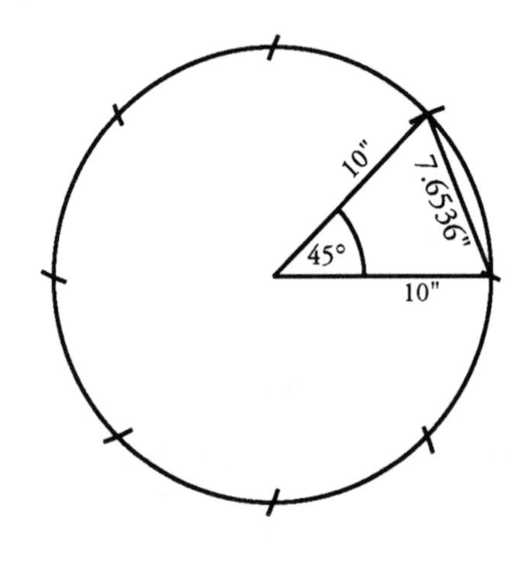

To mark chords, use a compass set at the chord length and mark points around the entire circle. Then take a straight edge and draw a line between the adjacent points. You can also use a straight ruler and repeatedly mark the chord length distance around the entire circle.

Chords of large circles are usually measured with tape measures. Whether you use a compass or a ruler, you are drawing straight lines which measure off equal chords, therefore equal sectors, of the circle.

Practice 67: Find the chord lengths for each problem. Leave the answers in decimal form. Answers-Page 222.

	Radius	# sectors		Radius	# sectors
(1)	12'	8	(4)	97.538"	9
(2)	5"	5	(5)	3 miles	16
(3)	100 yds	12	(6)	$22\frac{9}{16}$"	6

notes

Regular Polygons

Polygons are many-sided geometric figures that enclose a space. We name polygons by the number of sides they have. A five-sided polygon is called a pentagon, while an eight-sided polygon is called an octagon. The least number of sides a polygon can have is three. (Three sided figures that enclose a space are called triangles.)

Of particular interest to us are regular polygons. **Regular polygons have equal angles and equal sides** and can be inscribed in a circle. Look at this octagon.

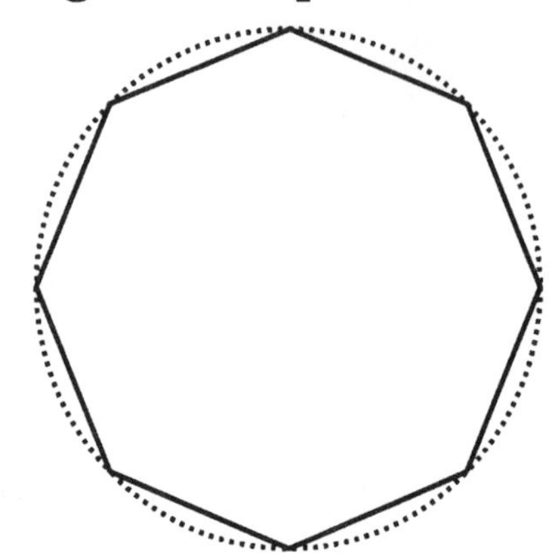

It should look familiar because it is the same shape as the one you just worked with on the previous page. The drawing on the previous page is a circle divided into eight equal sectors with the chords drawn in.

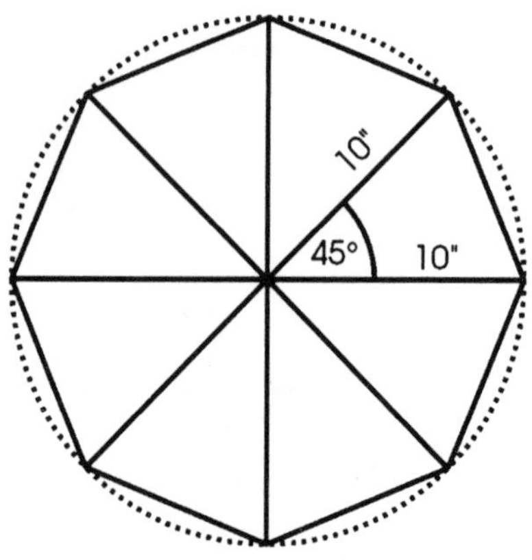

All regular polygons are created by dividing a circle into a number of equal sectors and adding chords. The number of sides of a regular polygon is also the number of sectors in the circle.

You can work any regular polygon by knowing the number of sides it has and either the length of the radius or the length of a side. When working with equal sectors and chords, you may not think of the figures you are working with as regular polygons, but that is exactly what they are.

On the next page is a list of the names of nine regular polygons. You won't see these names often, but when they do come up you will know that you can work them as equal sectors and chords of circles.

Number of sides	Name of polygon
3	Triangle
4	Quadrilateral
5	Pentagon
6	Hexagon
7	Heptagon
8	Octagon
9	Nonagon
10	Decagon
12	Dodecagon

Remember: **Regular polygons have equal angles and equal sides.**

For a triangle to be a regular polygon, it must be an equilateral and equiangular triangle. Those are fancy names for a triangle with equal sides and equal angles.

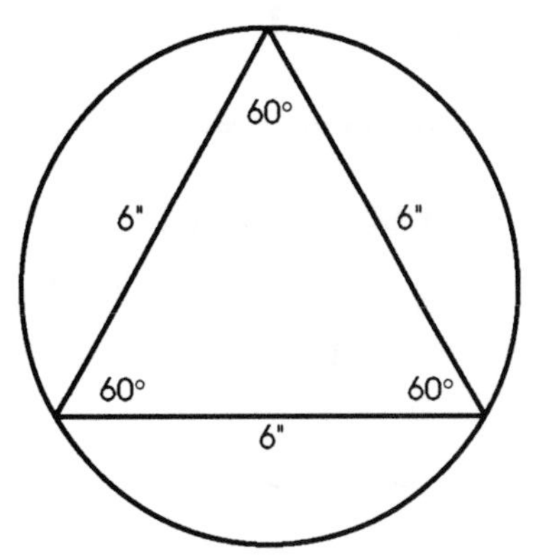

The only regular polygon for the four sided figure—the quadrilateral—is a square. The square has four equal sides and four 90° angles.

Practice 68: Calculate the length of one of the sides of each of these regular polygons. Answers-Page 222.

	Radius	Polygon		Radius	Polygon
(1)	20'	Hexagon	(5)	18"	Heptagon
(2)	22 miles	Dodecagon	(6)	6"	Dodecagon
(3)	27.4'	Decagon	(7)	16'	Quadrilateral
(4)	84"	Octagon	(8)	23.5625"	Pentagon

Remember: **When given the name of the regular polygon to work with, look up the number of sides, then treat the regular polygon as if it were inscribed in a circle with equal sectors and chords.**

Rolling Offsets

This special section is for people who work with pipe, conduit, and tubing. You were introduced to the *offset box* in the section on simple offsets. It was used to show how simple offsets can be drawn on the vertical and horizontal planes (sides) of the box.

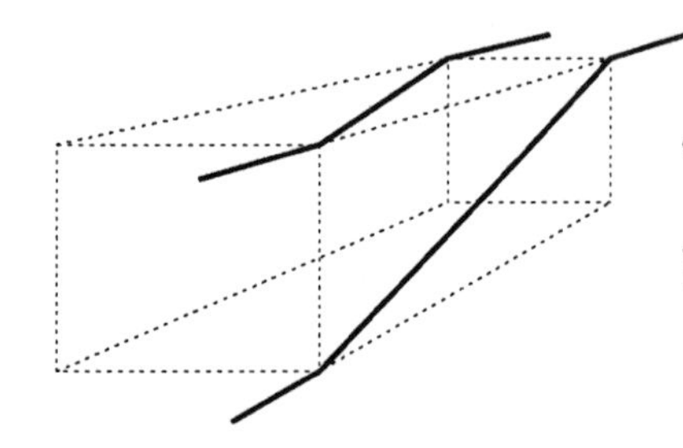

The simple offset can be drawn on the sides, top or bottom of the offset box.

Rolling offsets are simple offsets rolled to the side. The center lines for rolling offsets run diagonally through the offset box from one corner to the opposite corner. Notice how such a line is shown in the drawing.

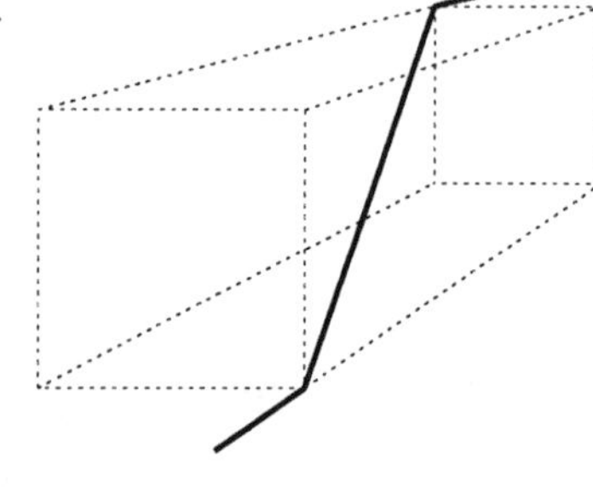

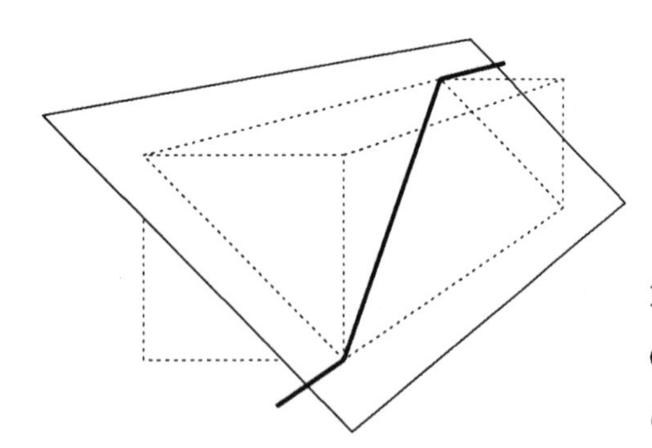

Just as with a simple offset, the center of a turn for a rolling offset is located in the corner of the offset box. Unlike with a simple offset, the *plane of a rolling offset bisects the offset box* instead of being a side of the box itself.

 *Remember:* **A diagonal line which bisects a rectangle creates two right triangles.**

Look at this drawing of a plane within the offset box. *The two right triangles which are placed diagonally in the box are the **offset triangles**.*

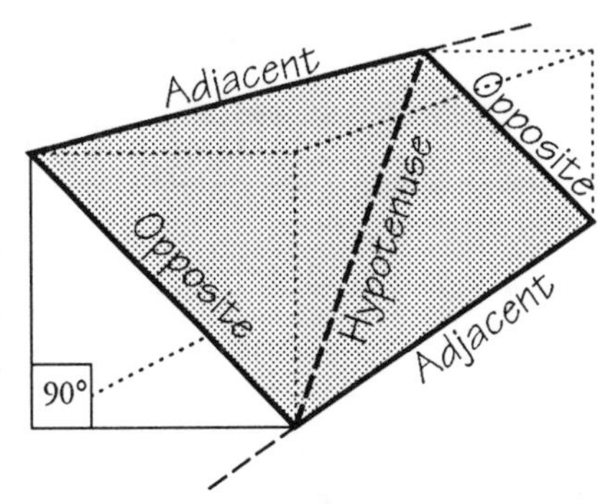

The names of the sides of the offset triangles have been placed in the drawing. Notice how the opposite sides of the offset triangles bisect the ends of the box and create more right triangles. The equal triangles at the ends of the box are called the **triangles of roll** and are the starting place for calculating rolling offsets.

The Triangle of Roll

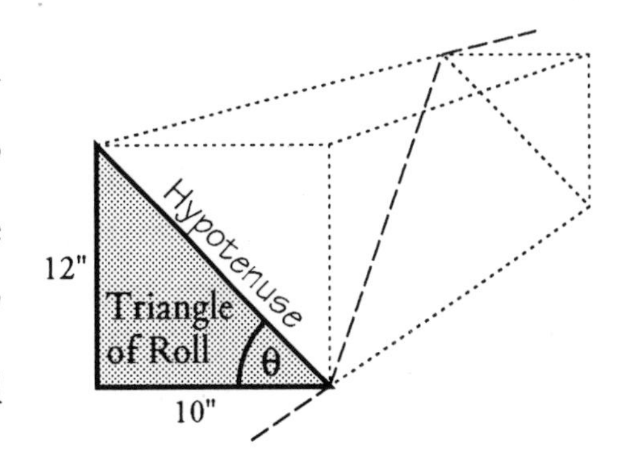

Two dimensions used in the triangle of roll which are essential to the rolling offset are the distance to be moved up (or down) *and* the distance to be moved over. If you need to move 10" over and 12" up, then the lengths of the legs of the triangle of roll are 10" and 12".

Notice that the *hypotenuse* of the triangle of roll is also the *opposite side* of the offset triangle. Once you have found the length of the hypotenuse for the triangle of roll, you can work the offset triangle as you would a simple offset.

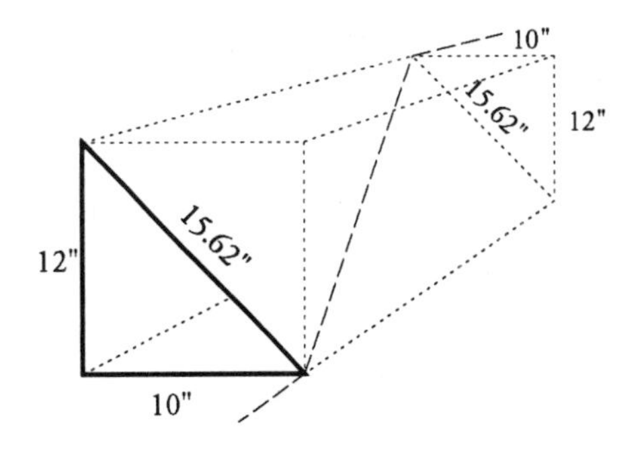

Look at the diagram to the right. Since two sides of the triangle are known, the length of the hypotenuse can be found by using the Pythagorean theorem.

$$c = \sqrt{a^2 + b^2}$$
$$c = \sqrt{10^2 + 12^2}$$
$$c = \sqrt{100 + 144}$$
$$c = \sqrt{244}$$
$$c = \mathbf{15.6205''}$$

The hypotenuse length for the triangle of roll *and* the length of the opposite side of the offset triangle is 15.6205".

Since the length of the opposite side of the offset is known, the offset triangle can now be worked as a simple offset. *The lengths of the hypotenuse and the length of the adjacent side of the offset triangle depend on the angle of turn used.*

The 45° Rolling Offset

If you choose a 45° angle of turn, you will work the offset triangle as you would any other 45° simple offset.

- The length of the adjacent side is the same as that of the opposite side.
- The length of the hypotenuse is found by multiplying the length of the opposite side by 1.4142.

$$1.4142 \times 15.6205 = 22.0905"$$

If you do not remember why the length of the adjacent side and the length of the opposite side are equal or why we are using 1.4142 as a multiplier, please turn to page 79 and review.

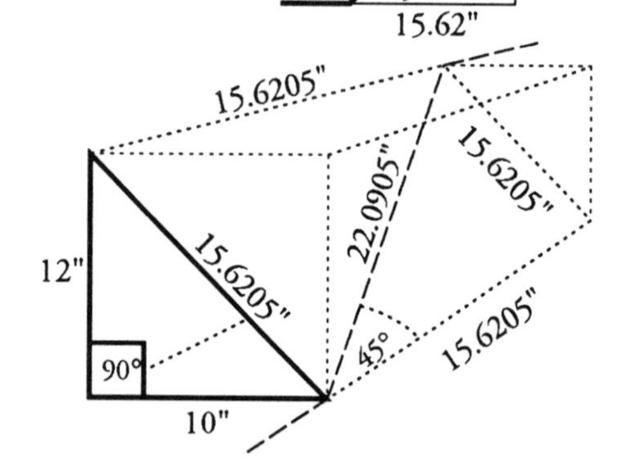

The 30° Rolling Offset

If you choose a 30° angle of turn, you will work the offset triangle as you would any other 30° simple offset.

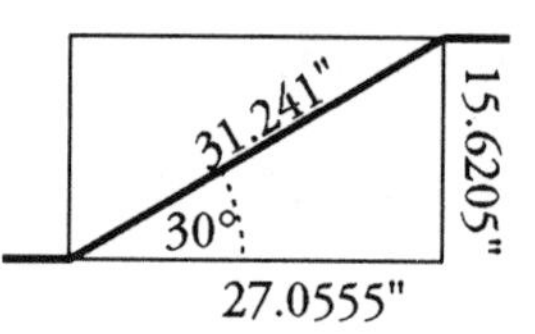

- The length of the adjacent side is found by multiplying $\sqrt{3}$ by the length of the opposite side.

$$1.732 \times 15.6205" = 27.0555"$$

- The length of the hypotenuse is found by multiplying the length of the opposite side by 2.

$$2 \times 15.6205" = 31.241"$$

If you do not remember why we are using 2 and $\sqrt{3}$ as multipliers, please turn to page 81 and review the section.

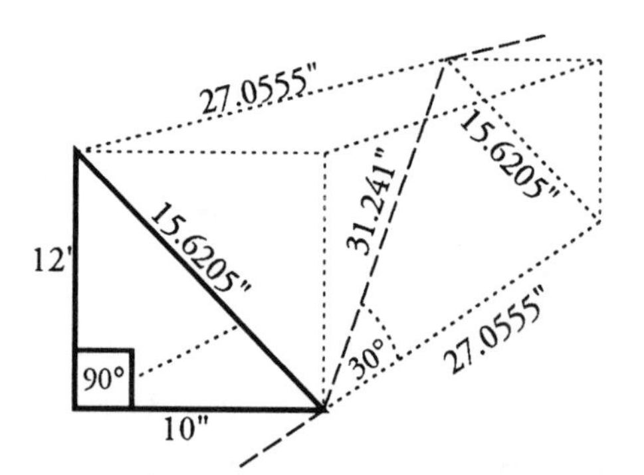

The Odd Angle Rolling Offset

If the distance a line can run is defined for you by the space available, you most likely will use an odd angle of turn.

Remember: The lengths of the hypotenuse and the length of the adjacent side of the offset triangle depend on the angle of turn used.

If you don't know the angle of turn to use, then you must find it before calculating the length of the hypotenuse of the offset triangle. For this example, the length of the adjacent side is limited to 20".

First: Find the angle of turn.

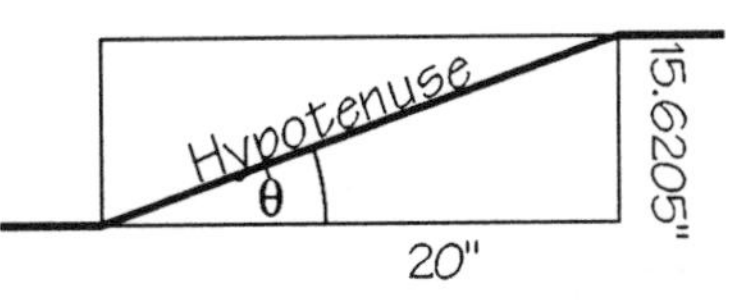

• By referring to the NAK chart, you will find that the angle of turn can be determined by using the tangent function.

$$\text{Tangent } \theta = \frac{\text{opposite side}}{\text{adjacent side}}$$

$$\text{Tangent } \theta = \frac{15.6205''}{20''}$$

$$\text{Tangent } \theta = .781025$$

$$\text{Tangent } \mathbf{38}^\circ = .781025$$

The angle of turn equals 38°.

Second: Find the length of the hypotenuse.

• By knowing two sides and one angle, you have a choice of two methods for finding the length of the hypotenuse. You can use the Pythagorean theorem or the NAK chart. We will use the Pythagorean theorem.

$$c = \sqrt{a^2 + b^2}$$

$$c = \sqrt{15.6205^2 + 20^2}$$

$$c = \sqrt{244 + 400}$$

$$c = \sqrt{644}$$

$$c = \mathbf{25.377''}$$

The 90° Rolling Offset

The 90° simple offset does not use an offset triangle; however, it still uses the triangle of roll. The length of the hypotenuse of the triangle of roll is the length needed for the offset. The drawing below shows what a 90° rolling offset looks like.

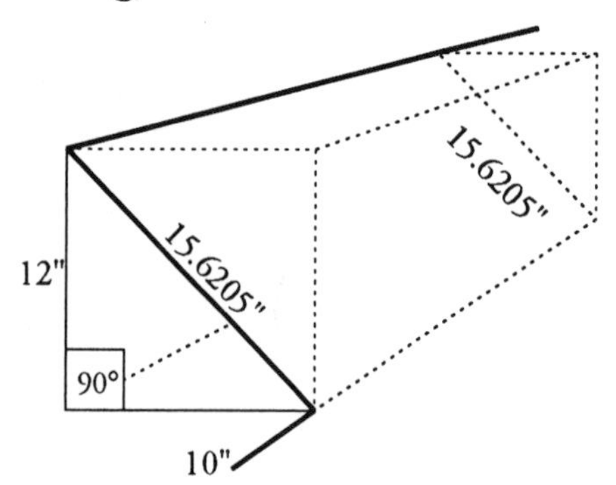

Steps for Working the Rolling Offset

Let's look at the steps for working a rolling offset.

First: Draw an offset box and identify the triangle of roll.

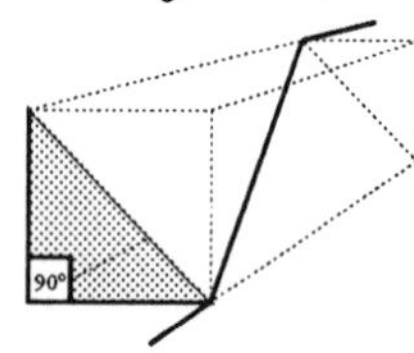

Second: Write in the known measurements.

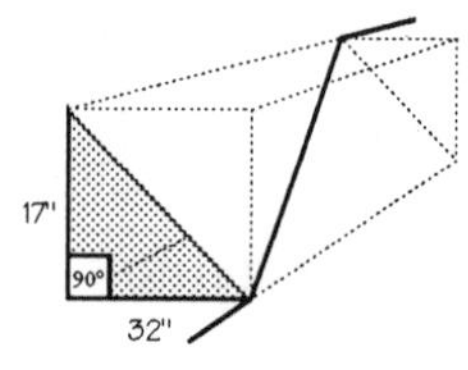

Third: Find the length of the hypotenuse of the triangle of roll using the Pythagorean Theorem.

$$c = \sqrt{a^2 + b^2}$$
$$c = \sqrt{17^2 + 32^2}$$
$$c = \sqrt{289 + 1024}$$
$$c = \sqrt{1313}$$
$$c = \mathbf{36.2353''}$$

If the rolling offset is a 90° rolling offset, your calculations are finished.

Remember: **The rolling offset is a simple offset rolled to the side. Once you have found the length of the hypotenuse of the triangle of roll, you work the rest as a simple offset.**

Fourth: Decide or calculate the angle of turn to use.

Note: If you need to calculate the angle of turn, the length of the adjacent side of the offset triangle needs to be known in order for you to use the tangent function.

$$\text{Tangent } \theta = \frac{\text{opposite side}}{\text{adjacent}}$$

Remember: **The length of the hypotenuse of the triangle of roll is also the opposite side of the offset triangle.**

Fifth: Find the length of the adjacent side and/or the length of the hypotenuse of the offset triangle.

For this example, use a 45° angle.

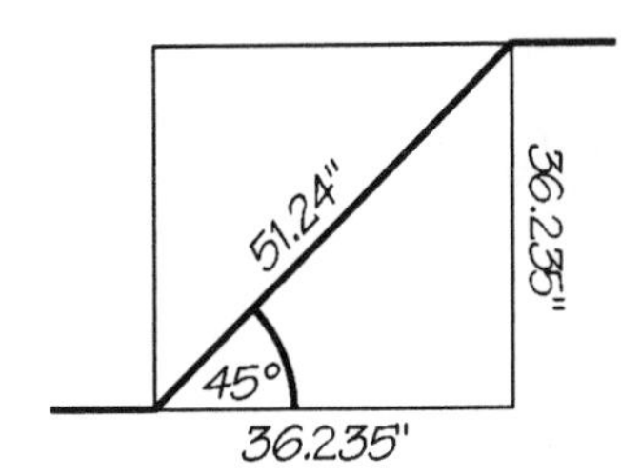

- The length of the adjacent side of a 45° simple offset is the same as the length of the opposite side.
- The length of the hypotenuse is found by multiplying the length of the opposite side by 1.4142.

36.2353" x 1.4142 = **51.24"**

Practice 69: Find the length of the hypotenuse of roll and the length of the hypotenuse of the offset triangle for these rolling offsets. Don't forget to draw the offset box to help you. Leave your answers in decimal form. Answers-Page 222.

	Over	Up	Angle of turn		Over	Up	Angle of turn
(1)	26"	19"	30°	(5)	22.75"	22.75"	45°
(2)	63"	37"	45°	(6)	22.75"	22.75"	90°
(3)	41'	54'	90°	(7)	43.563"	17"	30°
(4)	1'	3'	30°	(8)	3'	4'	adj. 6'

Notes about the Angle of Roll

There are cases when the angle of roll needs to be known. Such cases occur when you are prefabricating the rolling offset and another bend or elbow needs to be placed or bent on the same piece of material. In most instances, the angle of roll is the angle from the bottom of the offset box to the hypotenuse as shown below.

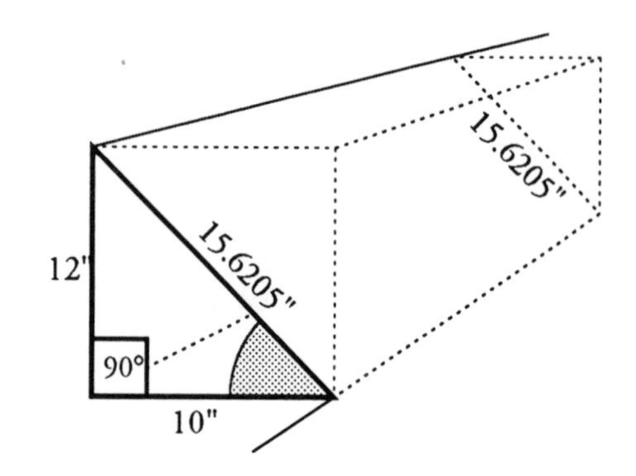

You can find that angle by using the tangent function.

$$\text{Tangent } \theta = \frac{\text{opposite side}}{\text{adjacent side}}$$

$$\text{Tangent } \theta = \frac{12''}{10''}$$

$$\text{Tangent } \theta = 1.2$$

$$\text{Tangent } \mathbf{50.2}^\circ = 1.2$$

This reference angle is used because it is the angle that the angle finder would read if the adjacent side of the triangle was ground level.

notes

XVI

The Metric System

We have all seen someone doing something the hard way and wanted to show them a much easier method. Yet, we ourselves do something the hard way whenever we use our measuring system. Ours is one of the most difficult systems in the world. In fact, the United States is the only major country still using this system based on the old British imperial system, one the British themselves abandon years ago in favor of metric. The other four countries that have not converted to the metric system are Burma, Brunei, Liberia, and Yemen.

Before the creation of the metric system, measuring systems were based on local customs. For the English, the inch was the length of three dry barley corns. The foot was the length of the king's foot. The yard was the length from the king's thumb to the tip of his nose. Of course, when a tall king died and a short king took his place, consistent measurements were made even more difficult. Not only did each country have its own system of measurements, but sometimes different sections of a country had different systems. To say the least, it was confusing.

About the time we were working on our constitution (the 1790's), the French National Assembly asked its country's scientists to create a standard system of measurements and weights. The new system had to be logical enough for all people to understand and accurate enough for scientists. The system they designed met those requirements. It is commonly called the metric system, although the official name is Systeme International d'Unites (International System of Units) or just SI .

In developing the metric system, the scientists first, assigned a basic unit to each of the seven major kinds of measurements. We will use the three in this list:

Measurement	Base Unit
Length	Meter
Weight or mass	Gram
Volume	Liter

The French scientist divided these base units to create fractions of units and multiplied them to created larger units. They decided to use the decimal system for all units and attach a **prefix** to the base units to indicate larger and smaller units.

For units smaller than the base unit, they added prefixes using *Latin* terms:

deci	meaning **ten** for	.1
centi	meaning **one hundred** for	.01
milli	meaning **one thousand** for	.001

For units larger than the base unit, they added prefixes using *Greek* terms:

deka	meaning **ten** for	10
hecto	meaning **one hundred** for	100
kilo	meaning **one thousand** for	1000

The development of the metric system created one unified method for identifying all measurements. Learn the basic units and prefixes, and the rest is easy.

kilometer	= 1000 meters	kilograms	= 1000 grams	kiloliters	= 1000 liters
hectometer	= 100 meters	hectograms	= 100 grams	hectoliters	= 100 liters
dekameters	= 10 meters	dekagrams	= 10 grams	dekaliters	= 10 liters
meters		**grams**		**liters**	
decimeters	= .1 meter	decigrams	= .1 gram	deciliters	= .1 liters
centimeters	= .01 meters	centigrams	= .01 grams	centiliters	= .01 liters
millimeters	= .001 meters	milligrams	= .001 grams	milliliters	= .001 liters

Length

If measurements are taken in metric, then the numbers from those measurements can be entered directly into a calculator. The answer is a unit of measurement that can be applied directly in the field. Here's an example of how to calculate a right triangle using metric measurements.

Find the length of the hypotenuse of a right triangle if a = 6 decimeters and b = 9 decimeters.

$$c = \sqrt{a^2 + b^2}$$

$$c = \sqrt{6^2 + 9^2} = \sqrt{36 + 81}$$

$$c = \sqrt{117}$$

$$c = 10.816 \text{ decimeters}$$

That's it. Converting isn't necessary. Using the metric system is easier than using the English system since you do not have any English measurements from the field to convert to decimal numbers on the calculator.

You can easily express the answer to the above problem in other metric units, such as meters, centimeters, millimeters, decameters, hectometers, and kilometers. Look at this chart:

Kilometers	.0010816
Hectometers	.010816
Dekameters	.10816
Meters	1.0816
Decimeters	**10.816**
Centimeters	108.16
Millimeters	1081.6

All of these figures identify the same length. Notice the movement of the decimal point. It moves one decimal place to the left for the next largest unit and one to the right for the next smallest unit. That's all you have to do to convert from one metric unit to another. Pretty simple, wouldn't you say?

Just for fun, let's take the same length in our English measuring system and run it up and down our units of measurement.

10.816 decimeters equal:

$42\frac{9}{16}''$	or	42.5826	inches
$3'\,6\frac{9}{16}''$	or	3.5486	feet
$1\text{ yd }6\frac{9}{16}''$	or	1.1829	yards
$\frac{1\text{ yd }6\frac{9}{16}''}{1760\text{ yds}}$ or $\frac{3'\,6\frac{9}{16}''}{5280'}$	or	.0006721	miles

Even when we use decimals, the English system is confusing. Each unit of measurement uses a different numbering system. A mile measures 1760 yards or 5280 feet. A yard measures 3 feet or 36 inches. A foot measures 12 inches. Inches can be broken down into any number of parts, including halves, fourths, eighths, sixteenths and thirty-seconds.

The metric system will finally be completely adopted in this country because it just makes good sense. We are stubborn people, but good sense usually wins us over.

METRIC - ENGLISH CONVERSION CHART

METRIC TO ENGLISH

TO CONVERT (A)	TO (B)	MULTIPLY (A) BY
Centimeters	Inches	0.3937
Cubic Centimeters	Cubic Inches	0.06102
Cubic Meter	Cubic Feet	35.31
Hectares	Acre	2.471
Kilograms	Pounds	2.205
Kilometers	Miles	0.6214
Liters	Gallons	0.2642
Meters	Feet	3.281
Meters	Yards	1.094
Square Centimeters	Square Inches	0.1550
Square Kilometers	Square Miles	0.3861
Square Meters	Square Feet	10.76
Square Meters	Square Yards	1.196

ENGLISH TO METRIC

TO CONVERT (A)	TO (B)	MULTIPLY (A) BY
Acres	Hectares	0.4047
Cubic Feet	Cubic Meters	0.02832
Cubic Inches	Cubic Centimeters	16.39
Feet	Meters	0.3048
Gallons	Liters	3.785
Inches	Centimeter	2.540
Miles	Kilometer	1.609
Pounds	Kilograms	0.4536
Square Feet	Square Meters	0.0929
Square Inches	Square Centimeters	6.452
Square Miles	Square Kilometers	2.590
Square Yards	Square Meters	0.8361
Yards	Meters	0.9144

notes

XVII

Seeing is Believing

Many people's understanding of math is enhanced by seeing it work. If you are one of those people, the following section is written for you. Working this section is not only helpful in developing your drawing skills, but in confirming calculations as you work in the other sections of this book.

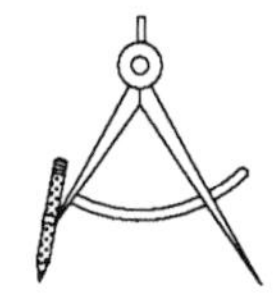

A compass icon has been placed throughout this book. It indicates that there is a drawing practice in this section to help with your understanding of that particular material.

The tools needed for this section are:

A **straight edge ruler** to make straight lines.

A **compass** (or trammel points) to make curved lines.

A **protractor** to visually check the angles.

Scissors or an **Exacto knife** to cut out the designs you draw

A **flexible ruler** (I use a $\frac{1}{4}$" wide x 8').

The materials needed are:

Newsprint (You can find it in large pads at art or office supply stores or go by your local newspaper and buy end rolls.)

Stiff cardboard, foam board, or **the sides of a box**.

Remember: Whether you use these tools or not, it always helps to draw problems out.

Fractions of the Ruler

A This practice should help in your understanding of why measuring fractions are based on the number two. We will start by showing how to divide a line into equal parts.

First: Draw a 6 inch straight line. Be very precise when measuring this line.

Second: Set your compass for a distance *a little longer* than one half of the length of that line (greater than 3 inches).

To *set* a compass means to adjust the radius of the compass to a certain measurement. A compass is set by placing the compass ruler on the sharp metal point side at zero and pulling the pencil point side out to the needed distance. (In this case it doesn't really matter how much past the halfway point you set the radius, just stay on the paper.)

Third: Place the metal end point of the compass on one of the end points of the line. Draw an arc (part of a circle, not a whole) through the line. Imagine where the center point of the line would be if it was placed directly above and below the horizontal line. Extend the arc so that it passes beyond where you imagine these points to be.

Fourth: Do not change the compass setting and mark the horizontal line from the other end point using the same method as above.

Fifth: Draw a straight line between the two points where the arcs cross.

There are now two line **segments**.

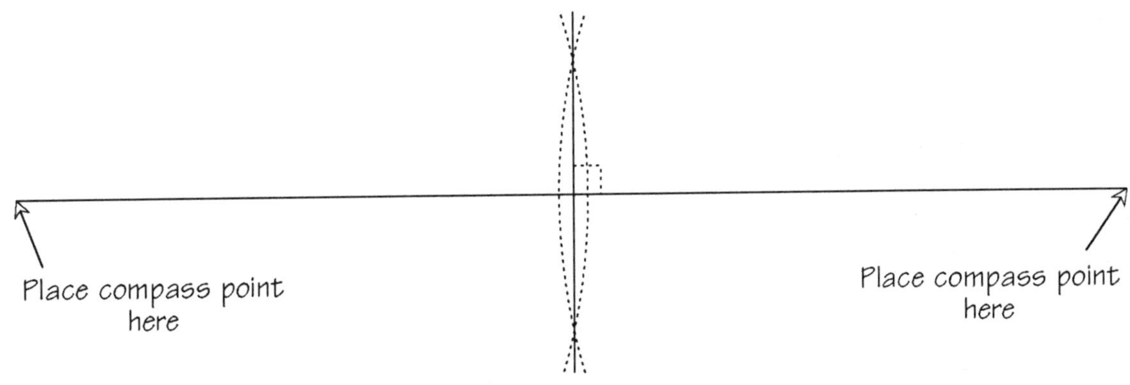

You accomplished two things with this action:

One: You divided the line exactly in half. See for yourself. Measure each side from the center point to the end point. Each segment should measure exactly three inches long.

Two: You created two lines that are **perpendicular** to each other.

You will study perpendicular lines soon, but what is significant now is that the line has been divided into equal segments. (Part of a line is called a **segment** of a line).

Since you know that the above action with a compass divides a straight line in half, let's repeat the same action on each of the two line segments just created.

First: Set the compass for *a little more* than one half of the distance between an end point and the center point (greater than 1.5 inches). Mark arcs through the line (as you did above) from each end point of each line segment.

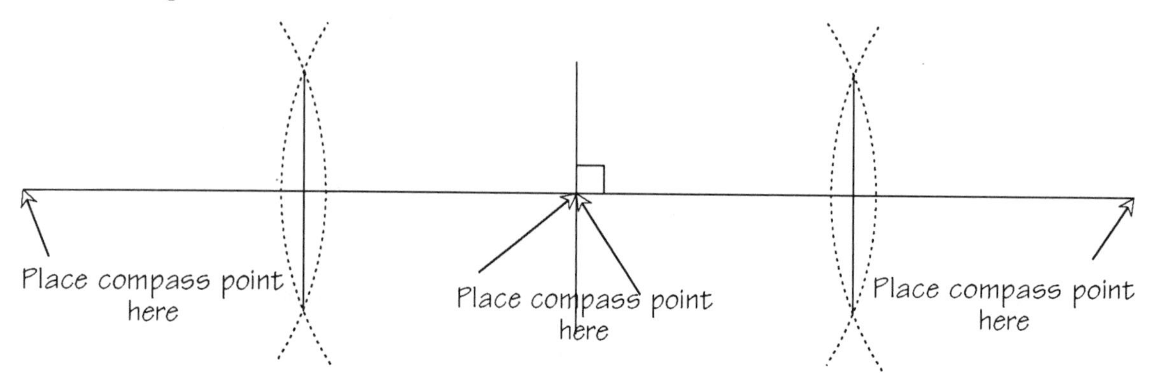

Second: Draw a straight line between the points where the arcs cross.

Each of the two segments have now been divided in half. The whole line has been divided into four equal parts or quarters. Measure the length of each segment. They should measure one and a half inches long.

If you again divide each of these line segments into equal parts, the line will be divided into eight equal parts (eighths) and each line segment will be three quarters of an inch in length.

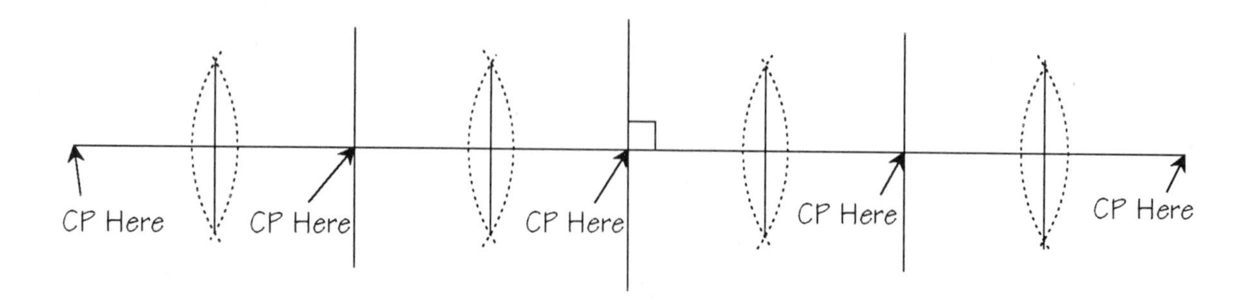

The next division will divide the whole line into sixteen equal parts.

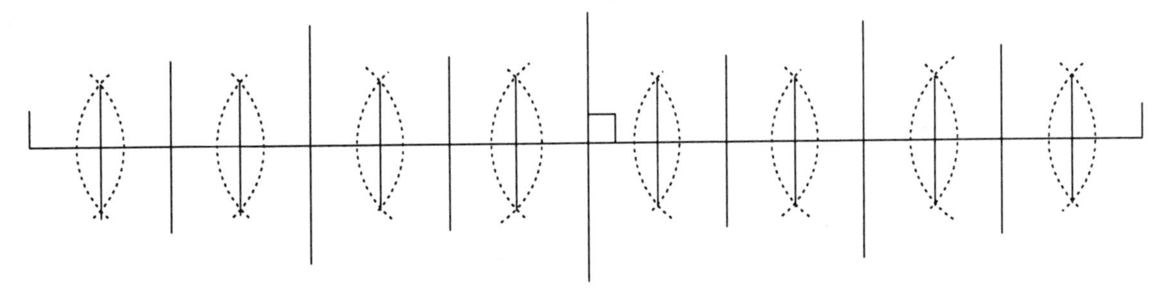

Each line segment should measure three eighths of an inch.

Turn back to page 5 and look at the fractions chart. Can you see why the English or customary measuring system divides inches into 2, 4, 8, 16, or more?

Perpendicular Lines

B This practice shows how to draw perpendicular lines. It was stated in the previous section that two perpendicular lines were created when a compass was used to divide a line into equal segments. If you want to draw a perpendicular from a particular point on a line, the process is the same; however, the approach is a little different.

First: Draw a straight line of any length and mark a point on the line.

Second: Using the same radius setting, place the metal end point of the compass on the marked point and mark the line on both sides of the point.

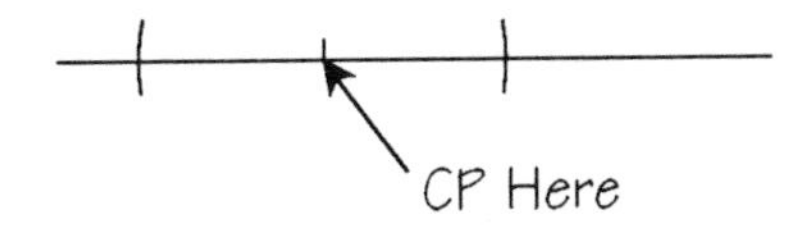

This makes the original point on the line a center point of the segment.

Third: Set your compass *a little longer* than the distance from one of the end points of the segment to the center point. Set your compass point on an endpoint of the segment and mark an arc above the center point.

Fourth: Keep the compass setting the same and repeat the action from the other end point.

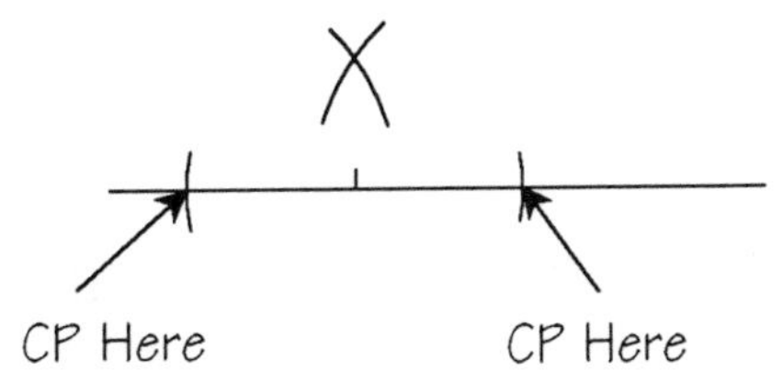

Fifth: Draw a straight line from the point where the arcs cross to the center point of the line.

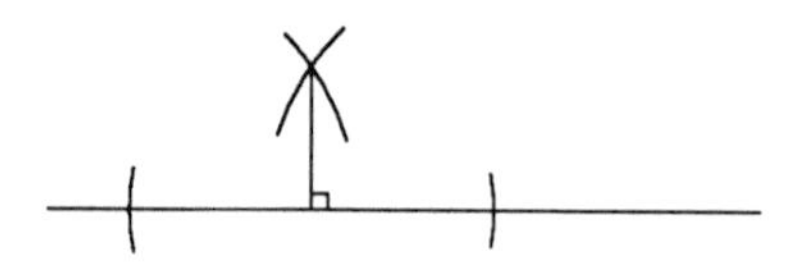

These two straight lines are perpendicular to each other.

Angles

C Angles can be equally divided using a straight edge and a compass.

First: Draw a straight line of any length and a perpendicular line to it (**B**). Notice that the angles created when you draw perpendicular lines are 90° angles.

Second: Set your compass for a distance less than the shortest line.

Third: Set the compass point at the vertex of the 90° angle and mark all lines. This action gives you three points which are the same distance from the vertex.

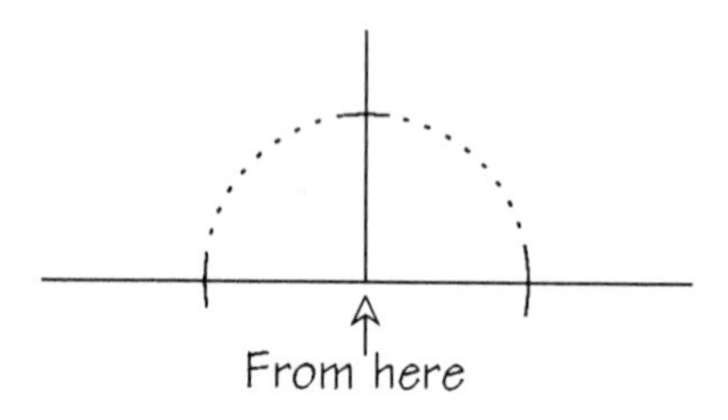

Fourth: Place the compass point at each of the three points and mark arcs between the lines.

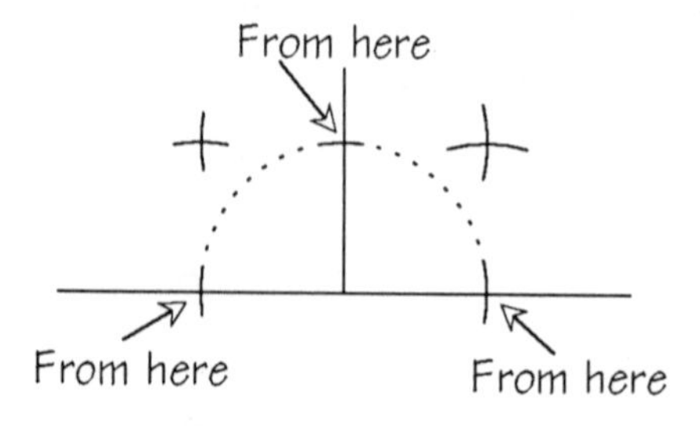

Fifth: Draw a straight line from each point where the arcs cross, to the vertex of the 90° angle. These lines equally divide the space between the 90° angles, making 45° angles.

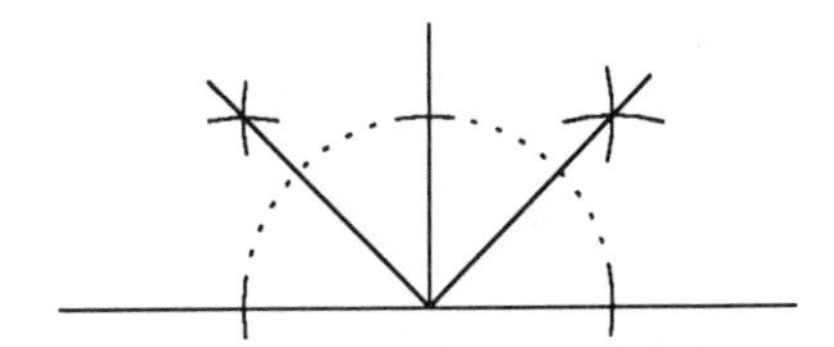

Sixth: Measure the angles with your protractor to confirm that the perpendicular lines have been bisected and that two 45° angles have been created.

Marking 60°, 30°, and 15° Angles

The first time I saw a compass, the kid next to me was drawing pretty designs like this with it. My mother got plenty of these whimsical paper flowers over the next few days. They were easy to draw since the compass never had to be reset. The same compass setting used to draw the circle was also used to mark the flower out. It was many years later before I realized that the same design was also the layout for dividing a circle into six equal parts.

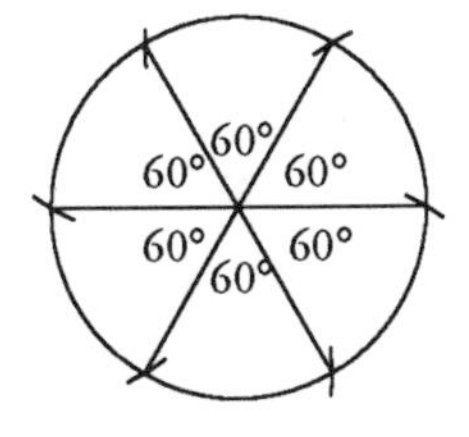

Since there are 360 degrees (°) in each circle, when a circle is divided into 6 equal parts, each part has 60°. $\frac{360°}{6} = 60°$

D This practice is designed to show how to mark 60°, 30° and 15° angles. You can mark any of these angles from a straight line using a compass and starting with a 60° angle.

First: Draw a straight line of any length and mark a point on the line close to the center. Set your compass point on the marked point and draw a semi-circle or half of a circle.

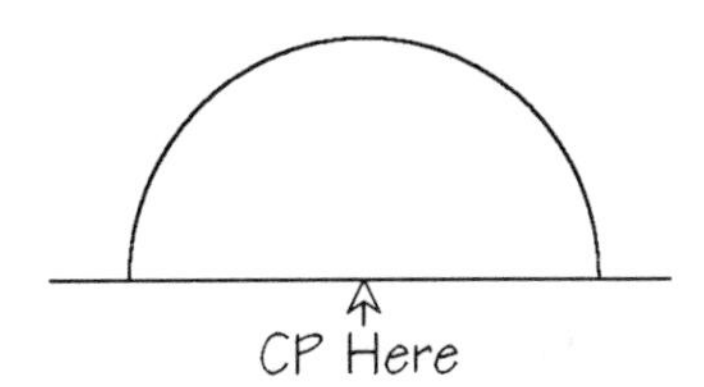

Second: Leave the setting of the compass the same as above and place the compass point where the semi-circle and the straight line meet. Mark a small arc across the semi-circle.

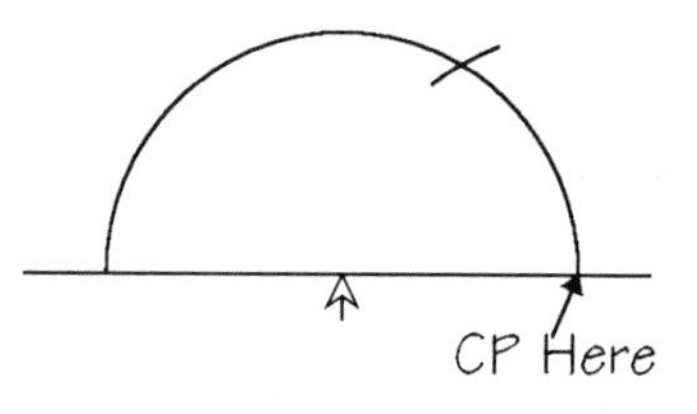

Third: Draw a line from the center point of your straight line to the mark on the semicircle. The line is 60° from the bottom line.

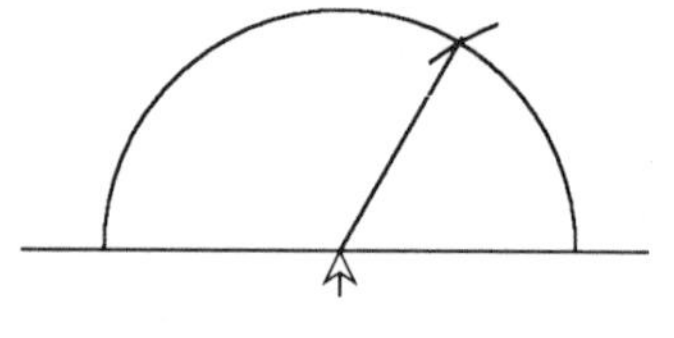

To mark a 30° angle, divide the sixty degree angle in half (*bisect the angle*). Since all points on the semi-circle are equal distance from the center, all you have to do to bisect the angle is:

First: Keep your compass setting the same and place the compass on the point where the semi-circle and one of the straight lines of the 60° angle meet. Mark an arc above, but within the boundaries of the 60° angle. Repeat the same action for the other 60° angle line.

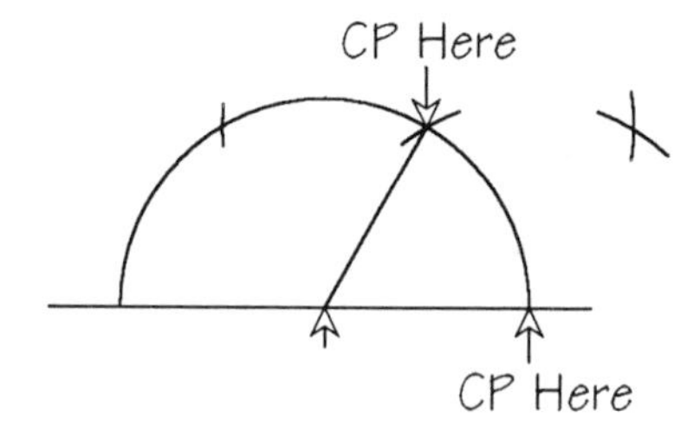

Second: Draw a straight line from the center point of the straight line (vertex of the 60° angle) to where the arcs cross. This causes the 60° angle to be bisected and results in two 30° angles.

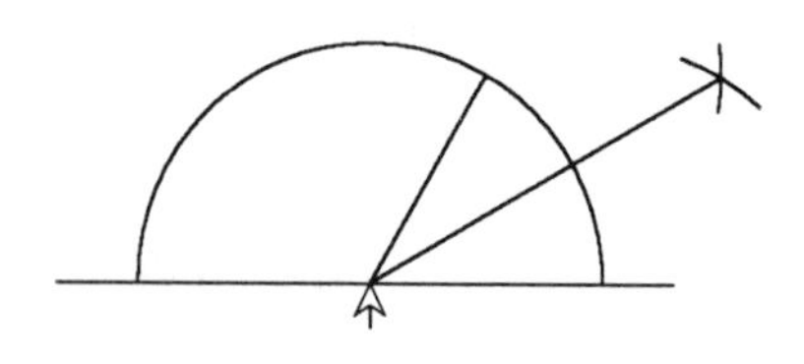

To mark a 15° angle, keep the compass setting the same and bisect one of the 30° angles.

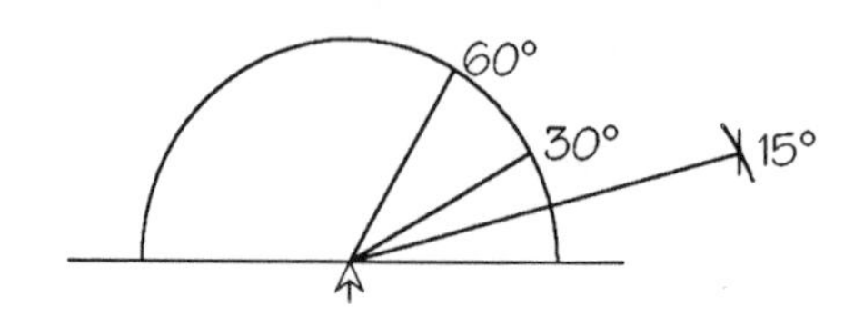

Right Triangles

E One of the most frequently drawn geometric shapes in the trades is the right triangle. Before you start working this practice which involves drawing right triangles, think about what makes a triangle a right triangle.

One of the three angles of a right triangle must be a **right** angle (90°).

Remember **Two straight lines that cross each other at right angles are perpendicular lines.**

First: Near the center of a piece of paper, draw a six inch straight line. Divide the line in half *and* create a perpendicular line through the center point. **(B)**. Make the perpendicular above the horizontal line 4" in length.

Since you divided the horizontal line in half, each horizontal line segment should measure three inches in length.

Second: Draw a straight line from the end point of the horizontal line to the endpoint of the upper vertical line.

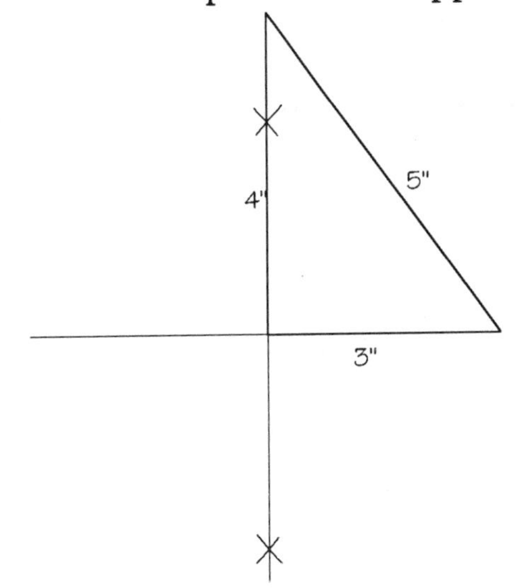

Measure the last line you drew. It should measure five inches. *If it does*, the triangle created is a 3-4-5 right triangle. *If it does not*, then either the length of one or both of the sides are not exactly three and four inches or the horizontal and vertical lines are not perpendicular.

Marking a 30°-60° Right Triangle

F *The shortest side of the 30°-60° right triangle is always half as long as the longest side* (hypotenuse). Follow the instructions below to draw a 30°-60° right triangle.

First: Use your knowledge of drawing and bisecting angles to draw a 30° angle (**D**).

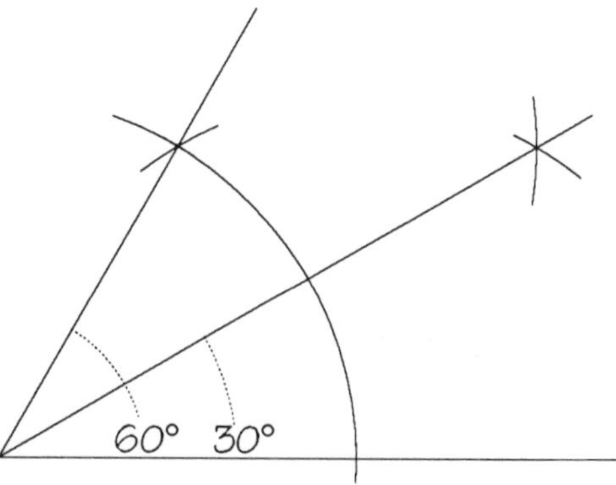

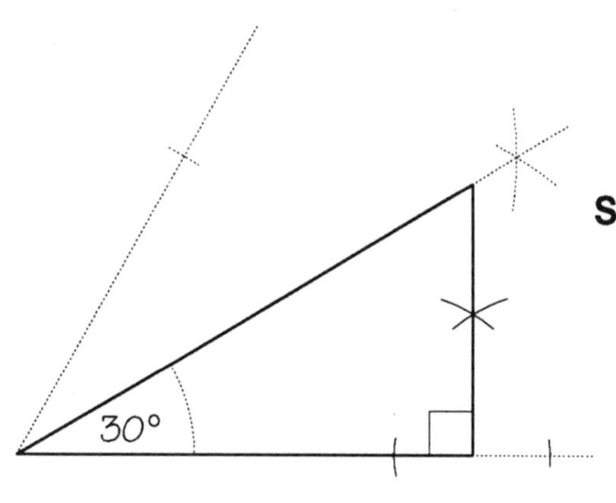

Second: From a point along the horizontal line, draw a perpendicular line that crosses the 30° line. Measure the shortest side (the perpendicular line) and the longest side (the hypotenuse) to confirm that the hypotenuse is twice as long as the shortest side.

If you are not convinced of the relationship between the longest and shortest side of a 30°-60° right triangle, draw another perpendicular line along the 30° line and measure that 30°-60° right triangle.

Marking 45° Right Triangles

G *Every 45° right triangle has two equal leg lengths and a hypotenuse length of 1.4142 times the length of one of those legs.* To verify this, follow the instructions below to first draw a 45° right triangle

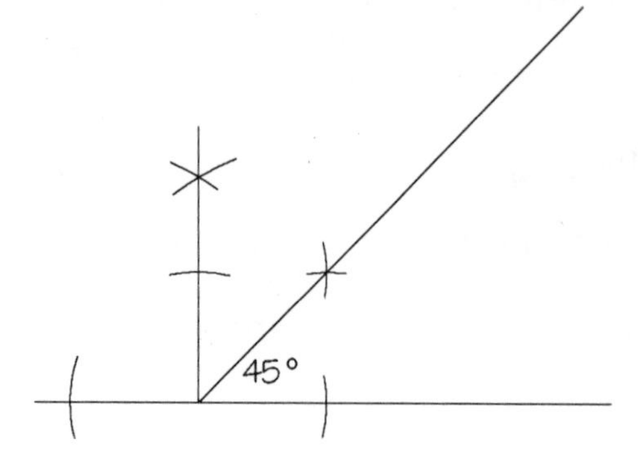

First: From a horizontal line, draw a perpendicular line (**B**), then bisect it to make a 45° angle (**C**).

Second: Draw two more perpendicular lines from the horizontal line so that each perpendicular line crosses the 45° line. Notice that each perpendicular creates a different 45° right triangle.

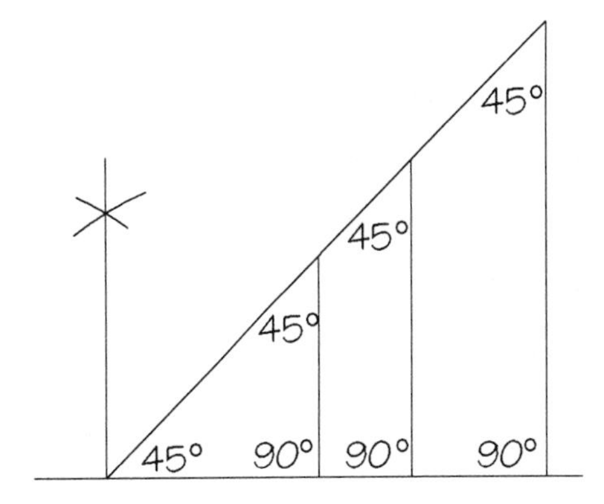

Third: Measure the length of each leg of each 45° right triangle.

For each triangle, there should be two equal legs and an hypotenuse with a length of 1.4142 times the length of a leg.

H Right triangles can be inscribed in a circle using the diameter of the circle as the hypotenuse for each triangle.

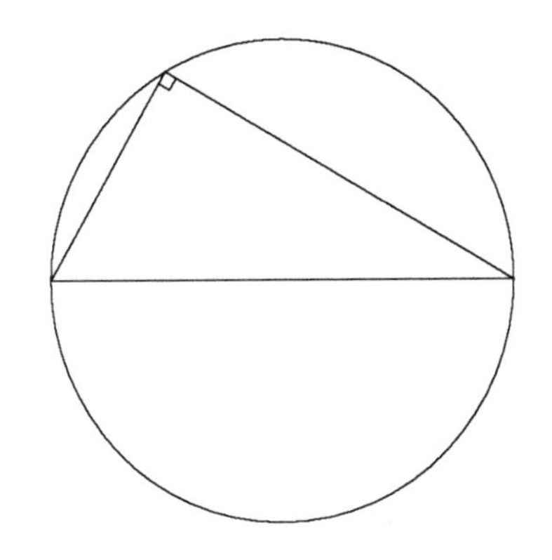

First: Draw any size circle and a diameter in that circle.

Second: Draw a straight line from the point where the circle and the diameter meet to any other point on the circle.

Third: Draw another straight line from the second point on the circle to the other endpoint of the diameter.

You have drawn a right triangle. Notice that the diameter is the hypotenuse of the right triangle.

Fourth: Draw four more right triangles in the same circle.

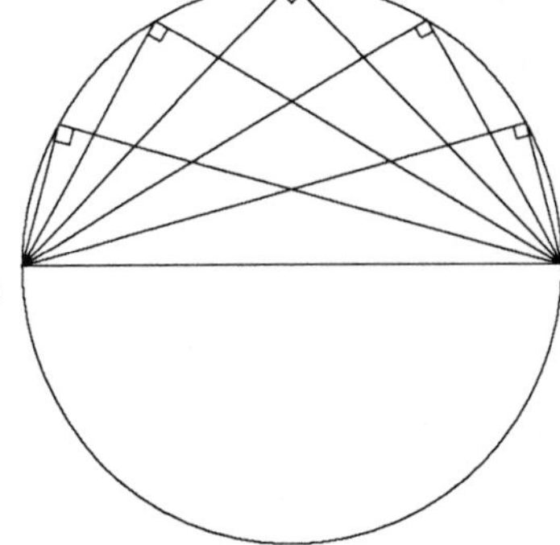

Measure the sides of each right triangle and notice that the hypotenuse is always the longest side.

Skilled carpenters use this knowledge to draw circles with their framing squares. They would set two nails with a distance between them equal to the diameter of a circle they needed. They would then place a framing square between the two nails and mark the heel of the square as they turned the square between the nails.

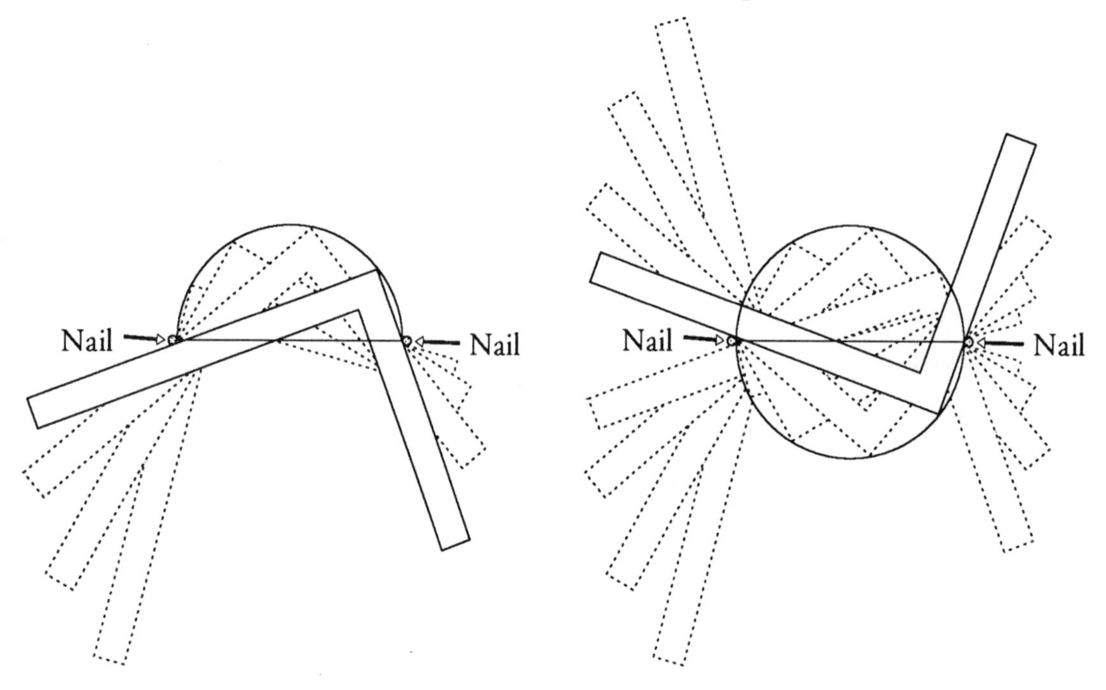

Notice that the diameter of the circle and the square create a right triangle at each position. Also notice that this procedure is not exactly accurate since the square never touches the center point of the nail. This is an example of accuracy needed on a job. Even though this technique is accurate enough for many rough carpentry jobs, there are many times in carpentry and other trades that this degree of accuracy is not acceptable.

Drawing Right Triangles Using Calculated Lengths

I Probably the most frequent means of drawing a right triangle is when the length of the sides of the right triangle have been calculated, and those dimensions are used to lay out the triangle.

For instance, you know a 45° right triangle with a leg length of 4" is needed. You also know that the legs of a 45° right triangle are perpendicular and are equal in length.

This knowledge is used in drawing that triangle.

First: Use a framing square or a compass to draw perpendicular lines. **(B)**

Second: Set the dimension of your compass for 4" and mark both lines from the vertex of the 90° angle. **(C)**

Third: Connect the points and you will have drawn a 45° right triangle.

For any right triangle, *if you know the lengths of the legs, you can draw perpendicular lines, measure and mark the needed distance for each leg from the vertex of the angle and connect the marks.*

Example: Draw a 26.5° right triangle with a opposite side length of 6".

In order to draw this triangle, the length of the adjacent side needs to be known.

First: Calculate for the length of the adjacent side (page 71).

A look at the NAK chart shows that if your *knowns* are an angle and the length of the opposite side and your need is the length of the adjacent side, the cotangent function is used for your calculation.

Cot 26.5° x the length of the opposite side = The length of the adjacent side
Cot 26.5° x 6" = The length of the adjacent side
2.006 x 6" = The length of the adjacent side
12.036" or $12\frac{1}{32}$" = The length of the adjacent side

Second: Draw perpendicular lines using a compass or a framing square **(B)**.

Third: Measure and mark $12\frac{1}{32}$" from the vertex of the angle on the horizontal line and 6" on the vertical line.

Fourth: Connect the two marks.

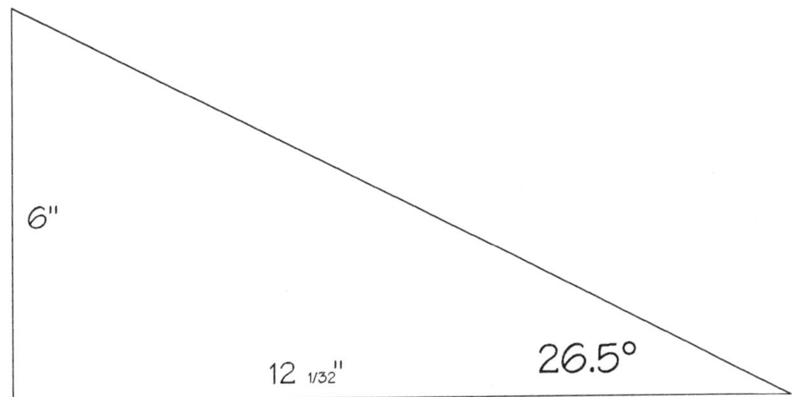

Protractors don't generally offer a high degrees of accuracy. If accuracy in your work is critical, then the ability to calculate and mark lengths for the sides of right triangles is also critical. For instance, If the length of the adjacent side was 12" instead of $12\frac{1}{32}$", the angle would be 26.6°.

For some of our work, a tenth of a degree makes little difference, but in other cases it does. Be aware and make calculations based on the accuracy needed for your work.

Finding Degrees On Sloped Objects

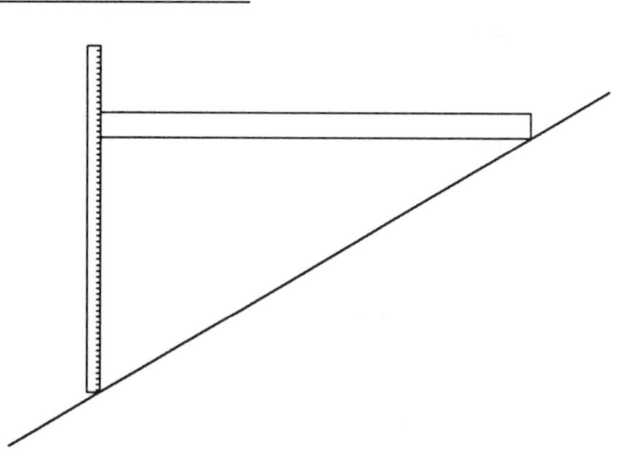

J Once you know how to calculate measurements for angles and leg lengths of right triangles, you can use that same knowledge to find the angle measurements for slopes. One leg length can be determined by holding a straightedge level to the slope (this drawing shows the straightedge placed horizontally) and measuring the distance between the other end of the straightedge and the slope. Since you know the length of the straightedge, you know the length of the other leg.

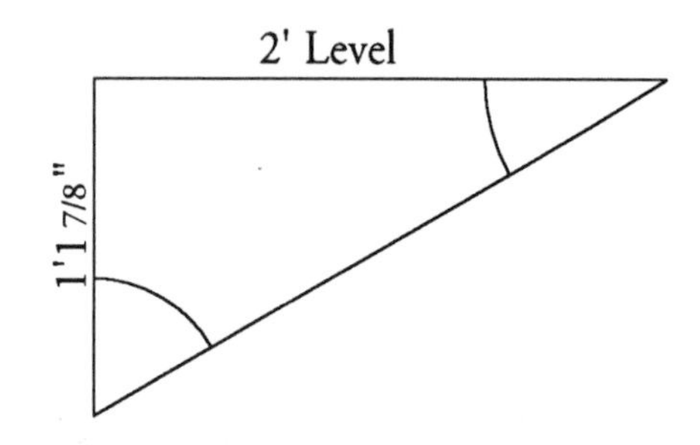

Once you have these measurements, calculating the angle is a matter of using the appropriate formula. Notice that no matter which angle you choose to calculate for, the known measurements are for the adjacent or the opposite side; however, which side is the adjacent side and which side is the opposite side is dependent upon which angle you are calculating for. Because of this, always be mindful of which of the two angles you are calculating for when using this procedure. Since the length of the adjacent side and the opposite side are known, the tangent function is used to find the angle.

Note: For the straightedge, I usually use a framing square with a torpedo level or a four foot level. I find this more accurate than using mechanical angle finders.

Isosceles Triangle

K Knowing that *a bisecting line divides an isosceles triangle into two equal right triangles* helps when drawing an isosceles triangle.

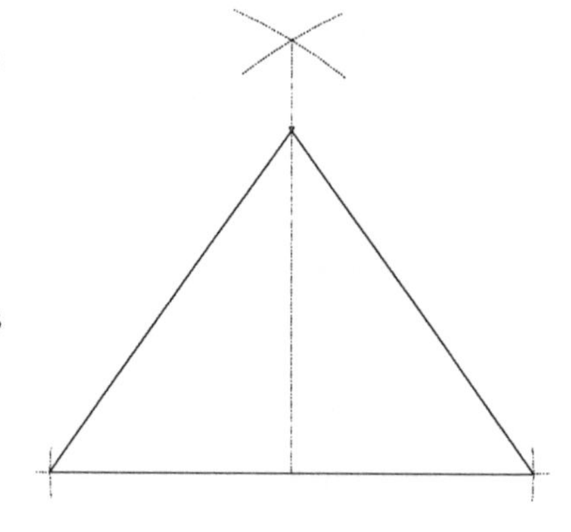

First: From a horizontal line, draw a perpendicular line using a compass.(**B**)

Second: Use your compass to mark equal line segments on both sides of the vertex of the 90° angles. (**A**).

Third: Pick a point on the vertical line and draw lines from both endpoints of the line segments to the point on the vertical line. *Voilà, an isosceles triangle.*

Circles

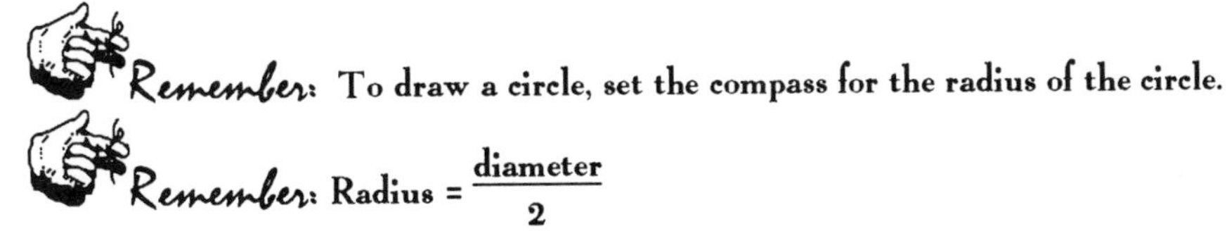

When drawing circles with a compass, the metal point end is placed at the center point of the circle and the other point (generally a pencil lead) is placed on the paper. The pencil point end of the compass is turned around until it gets back to the starting place.

For larger circles, trammel points can be used. Trammel points are devices that can be attached to straightedges to make circles. One trammel point has a metal point and the other has a pencil lead or pen tip. Trammel points are used just like a compass.

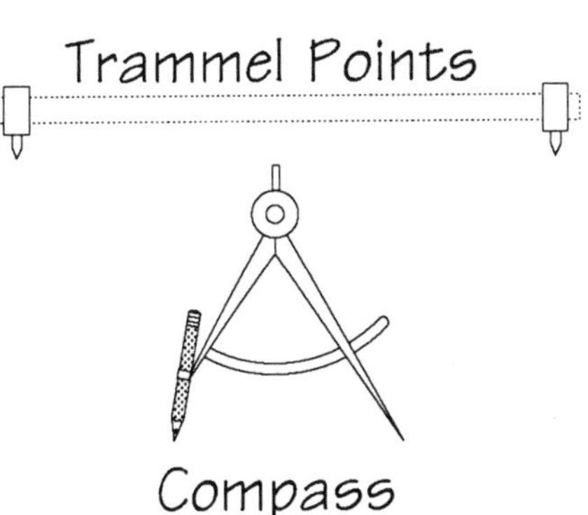

L In this exercise, you are going to cut circles out so that the circumferences can be rolled to find their lengths. For you to be successful with this practice, it is necessary that the compass or trammel points used to draw these circles maintains its accuracy. The ideal tool is a compass or trammel points which has a razor for an end point, instead of a pencil lead or pen tip; However, I have not been able to find this tool on the market. If you can find one or rig up one which will maintain the accuracy needed, your measurements will be more exact. If not, do your best with scissors or an Exacto knife and realize that there may be a slight deviation in measurements because of the tools used.

The tools and materials needed for this exercise are:

1. Trammel points or compass.
2. A piece of at least 9" square foam board, cardboard, etc. (Any material which can be cut by scissors or a razor and will hold its shape is OK.)
3. A straight edge ruler, scissors or razor, and a pencil.
4. Something to draw a 3' line on.

First: Set your trammel points or compass for four inches.

Second: On a piece of foam board, stiff cardboard, or the side of a box, use the trammel points or compass to draw a circle.

Third: Cut the circle out.

Fourth: With your straightedge, (on a separate piece of paper, cardboard, or on the floor etc.) draw a straight line approximately three feet in length. Mark an end point for this line.

Fifth: On the circle you've just cut, mark a starting point. Place the starting point of the circle *exactly* on top of the endpoint of the line. Roll the circle down the line until the starting point on the circle meets the line again. Mark the line at this point.

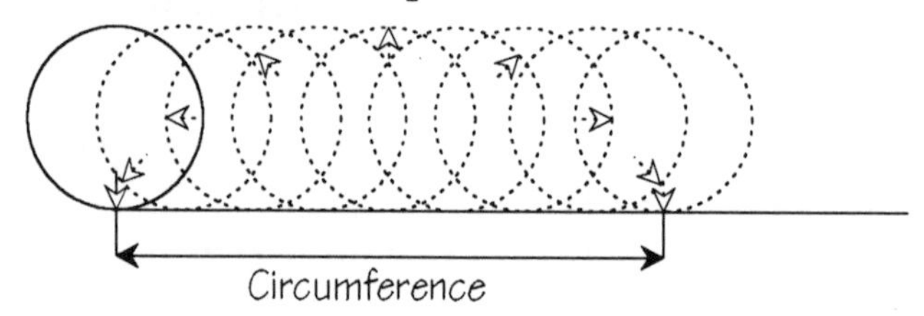

Sixth: Measure the distance between the two points. If the circle is *exactly* 8" in diameter (4" radius) and if you roll the circle without slipping, the distance is $25\frac{1}{8}$".

$$\text{Circumference} = \pi d$$
$$\text{Circumference} = 3.1416 \times 8''$$
$$\text{Circumference} = 25.1328'' \text{ or } 25\tfrac{1}{8}''$$

If the diameter of your circle is not exactly 8", multiply whatever the diameter of your circle is by π. This will give you the measurement for the circumference of your circle. If you aren't convinced that πd gives you the measurement of the circumference of any circle, draw, cut, roll out and measure more circles with different diameter measurements. Quit when you are convinced that πd is the correct formula for the circumference of a circle.

Arcs

M In this practice, you will divide a circle into different sectors and verify arc length for each sector.

Remember: An arc is a fraction of a circle.

First: Draw a four inch diameter circle.

Second: Draw in a diameter (**G**).

Third: Draw a line perpendicular to the diameter using the center of the circle as a point on the perpendicular line (**B**). Extend this line to also make it a diameter.

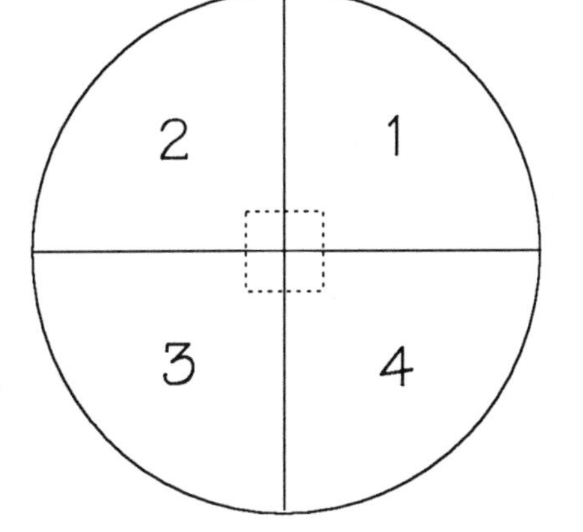

These actions divide the circle into four quadrants. (A quadrant is one fourth of a circle, a 90° sector.)

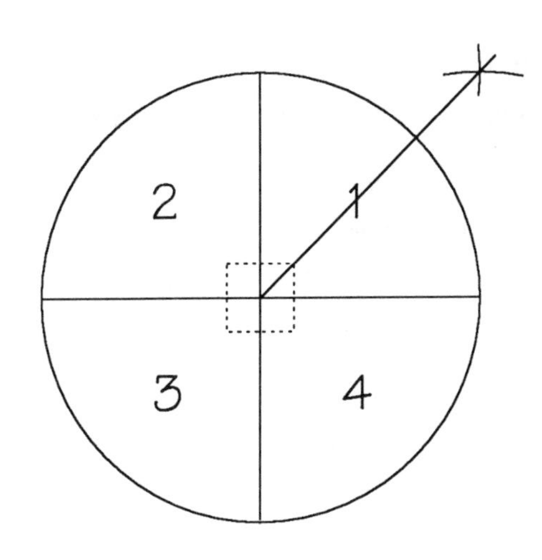

Fourth: Number the quadrants as shown. In quadrant 1, bisect the right angle to create two 45° arcs (**C**).

Fifth: In quadrant 2, use the radius of the circle as the compass setting and mark a 60° arc (**D**). Draw a line from the center of the circle to the mark on the circle.

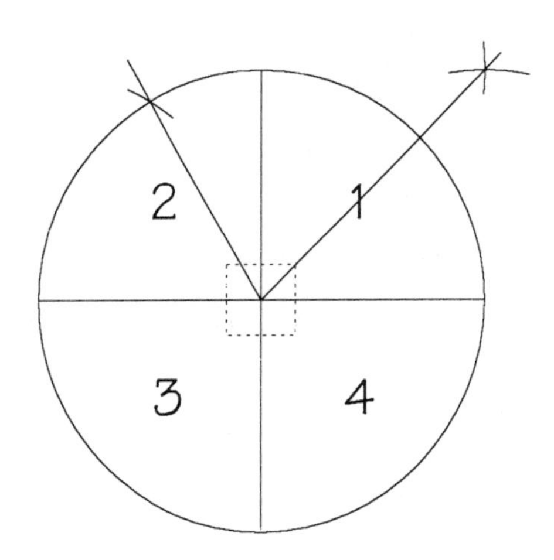

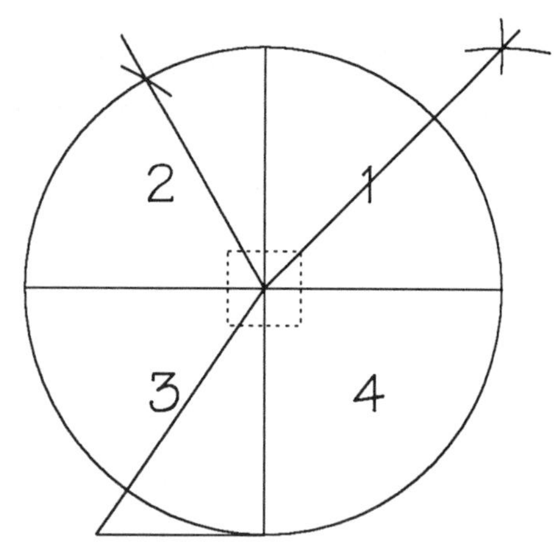

Sixth: Calculate the length of the opposite side of a right triangle which has a 40° reference angle and an adjacent side of 2" (length of the radius). In the third quadrant, use the line between quadrant 3 and 4 as the adjacent side of a triangle and draw in the opposite side using the measurement from the above calculation. Next, draw in the hypotenuse of the right triangle.

This drawing shows the different angles drawn in.

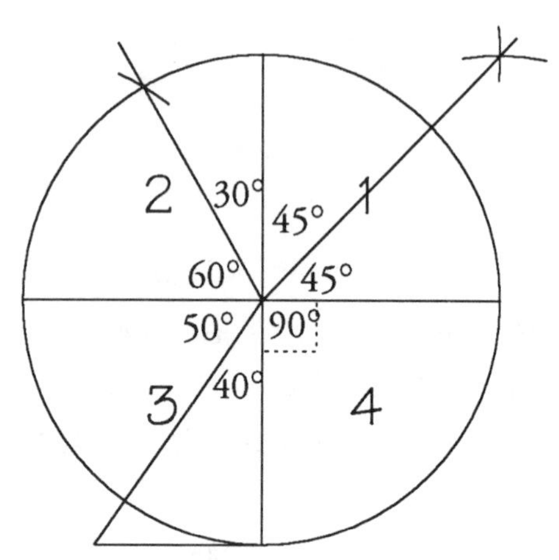

Seventh: Cut the circle out.

The circumference of the circle is found by using the formula πd. Roll out the circle to confirm that it is 6.2832".

Eighth: Now cut the circle into parts, using the lines you have drawn in as cutting guides.

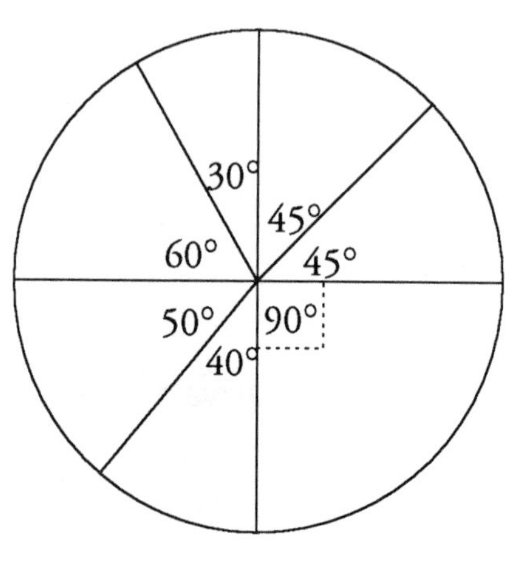

You have seven sectors: one 90°, two 45°, one 30°, one 40°, one 50°, and one 60°.

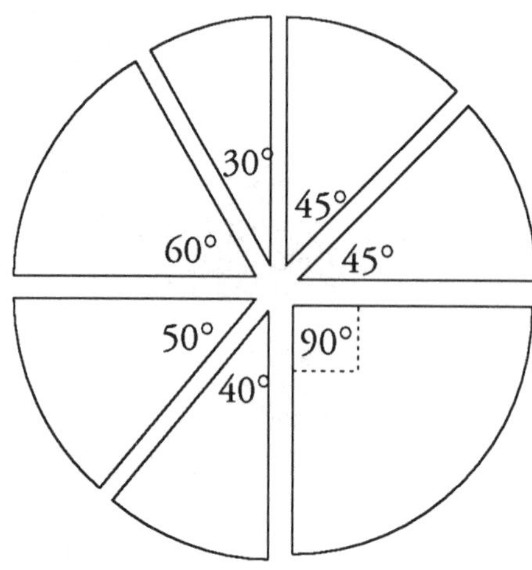

With these seven sectors you can confirm your calculations for arc length.

Ninth: Calculate arc length for each arc and confirm your calculations by rolling out and measuring each of the arcs.

Concentric Arcs

In this section, we're going to look at elbows and the calculations needed when elbows are joined to straight pieces of material.

 Let's first look at the 90° elbow.

First: Draw a four inch radius circle and divide it into quadrants (**M**).

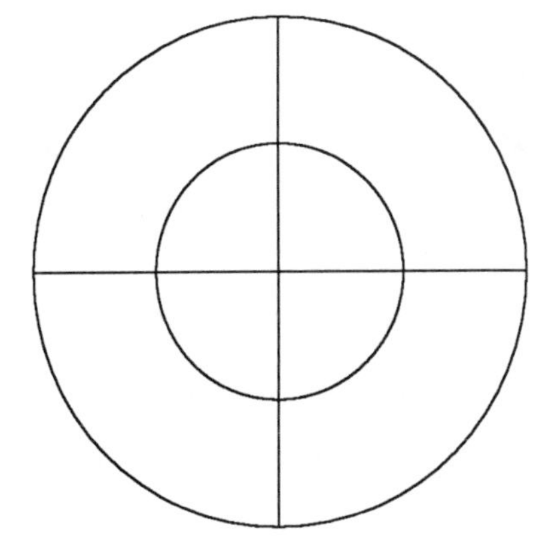

Second: Set your compass for a two inch radius and draw another circle, concentric (using the same center point) to the first circle (**N**).

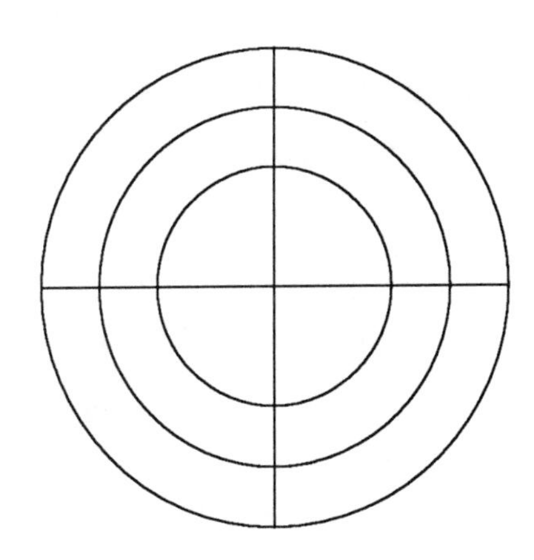

Third: Set your compass for a three inch radius and make another concentric circle exactly halfway between the other two circles. You now have three concentric circles. Their radii are 2", 3", and 4".

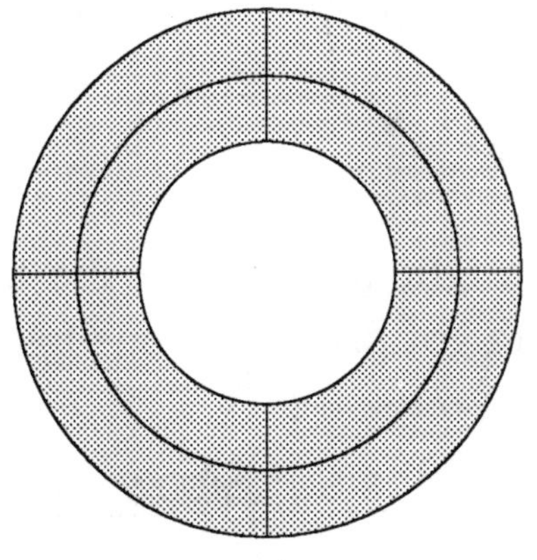

Fourth: Cut the circle out using the outside line as a guide. First, cut out the outside circle. Then cut the inside circle out.

The circle should looks like this.

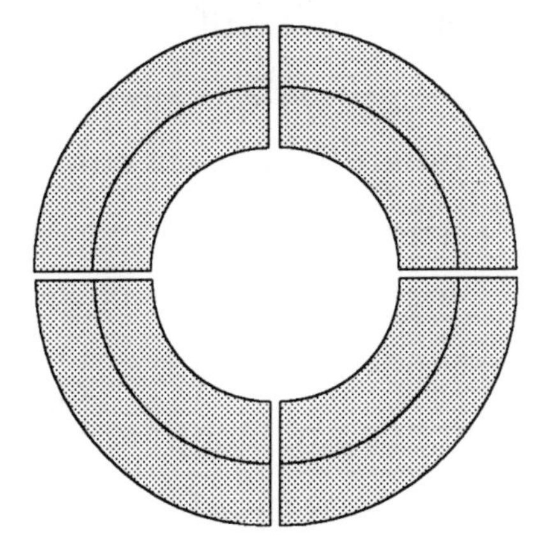

Fifth: Cut the quadrant lines to create four 90° elbows.

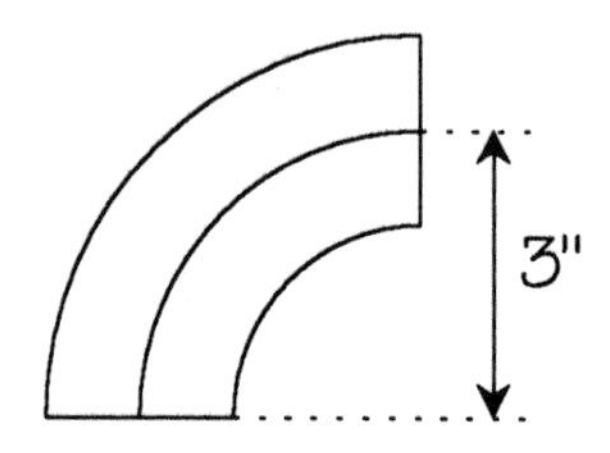

O The length of the take out for a 90° elbow is the same distance as the center radius measurement of the elbow. To check that statement: Place the bottom of the elbow on a flat surface and measure from the bottom face of the elbow to the center line on the upper face of the elbow.

Remember: Take out = $\tan \frac{\theta}{2}$ x radius of turn With 90° arcs, the radius of the arc and the take out of the arc are the same since the tangent of 45° is 1.

P Let's use these elbows in an offset.

First: Cut three straight pieces of cardboard or foam board 2" wide and 1' long. Be precise.

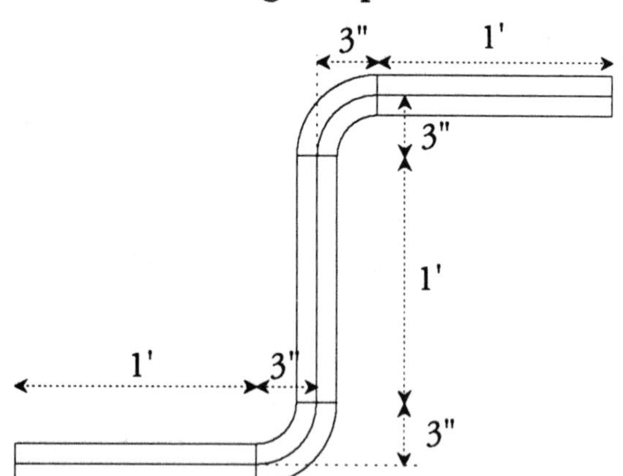

Second: Use the three straight pieces and two elbows (from **N**) to make an offset like the one shown on the left.

Third: Measure the offset from center line to center line.

The horizontal runs are 1'3" (1'+3") in length and the vertical run is 1'6" (1'+3"+ 3").

Notice that these lengths are 1' from each straight piece, plus the take outs of the elbows. The horizontal runs have one straight piece plus one take out and the vertical run has one straight piece plus two take outs.

Cutting the 90° Elbow Into 45° Elbows.

Q Let's cut one of these 90° elbows into smaller 45° elbows. A quick division of $\frac{90°}{45°}$ shows that two 45° elbows can be created from one 90° elbow. Below you will find three different ways to calculate measurements to mark 45° elbows.

The *first* method is: Measuring or calculate the inside and outside arc length for the 90° arc and divide these lengths by two.

90° outside arc length = 6.2832"

45° outside arc length = $\frac{6.2832"}{2}$ or 3.1416"

90° inside arc length = 3.1416"

$$45° \text{ inside arc length} = \frac{3.1416''}{2} \text{ or } 1.5708''$$

You can measure and mark the length of the outside arc by either rolling the arc or by using a flexible tape measure. The inside arc will have to be measured using a flexible tape measure.

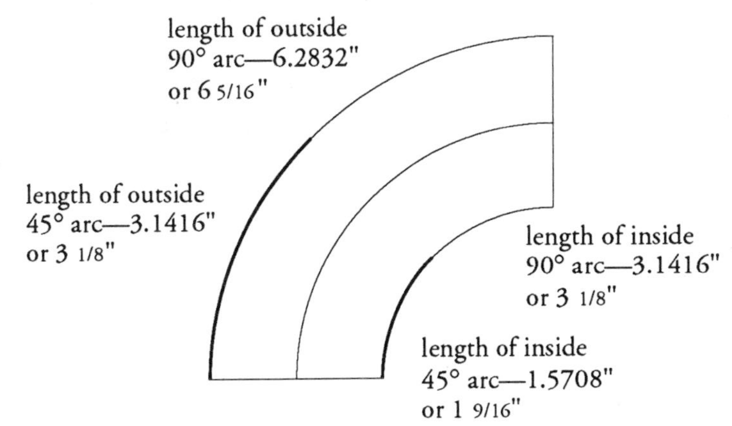

The *second* method is: Calculate the inside and outside arc lengths for a 45° elbow and mark and cut the 90° elbow from these calculations.

Arc length = radius x radian

Outside arc radius = **4"**

$45° = 45 \times \frac{\pi}{180} =$ **.7854 radians**

Outside 45° arc length = 4" x .7854

Outside 45° arc length = 3.1416"

Inside arc radius = **2"**

Inside 45° arc length = 2" x .7854

Inside 45° arc length = 1.5708"

The *third* method is: Find the length of the outside and inside arcs, divide these measurements by 90, and multiply the answers by the degrees (in this case, 45) of the desired elbow.

We have already measured the outside and inside arcs for the 90° elbow.
Outside arc length = 6.2832" and
Inside arc length = 3.1416".

Divide the 90° outside arc by 90, then multiply the answer by 45.

$$\frac{6.2832''}{90} = .06981333$$

.069813333 x 45 = **3.1416"**

Divide the 90° inside arc by 90, then multiply the answer by 45.

$$\frac{3.1416''}{90} = .034906666$$

.034906666 x 45 = **1.5708"**

Notice that all methods of calculation provide the same answer.

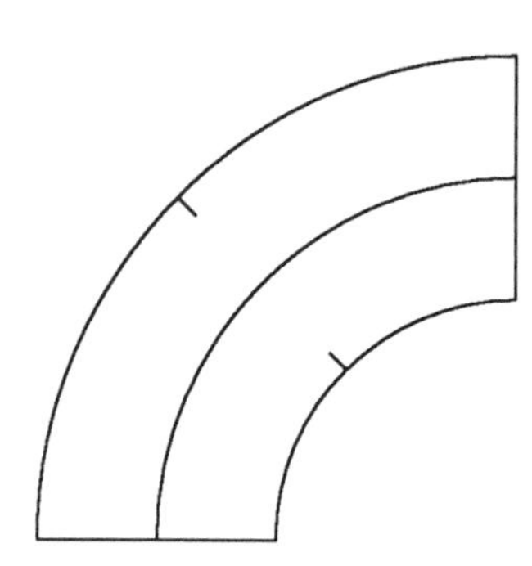

These calculations are used to mark the 90° elbow. Once the 45° arcs are marked on the 90° elbow, it can be cut to form two 45° elbows.

First: Use the measurements from one of the above calculations to mark the arcs for two 45° elbows.

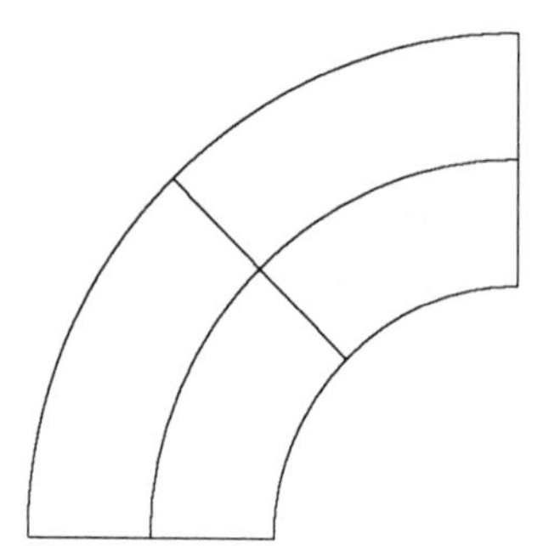

Connect the marks by drawing a line between the two marked points. Cut the elbow at the marked line to make two 45° elbows.

R Let's look closely at take outs for a 45° elbow.

Remember: The take out of an elbow or bend is calculated based on the center radius.

First: From the center arc on both faces of two 45° elbows, draw perpendicular lines toward the center of the elbow.

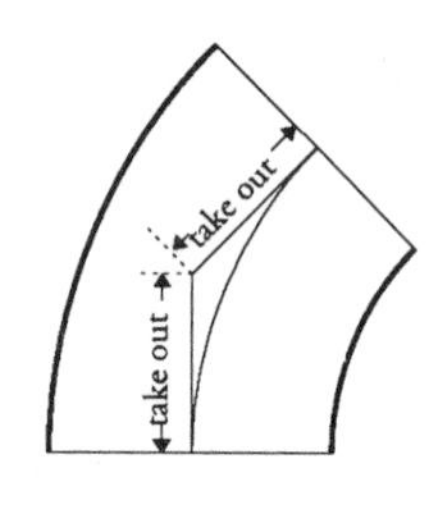

The distance from the face of the elbow to the point where the lines meet is the length of the take out.

Here are the calculations for take out for a 3" 45° arc.

Take out = $\tan\frac{\theta}{2}$ x radius of turn   Center radius = 3"

Take out = tan 22.5° x 3"   tan 22.5° = .4142

Take out = .4142 x 3"

Take out = 1.2426" ($1\frac{1}{4}$")

Second: Measure the lines you just drew in. They should measure $1\frac{1}{4}$", the same as the calculations

Third: Take the one foot straight pieces you cut for the previous exercise and use two 45° elbows to make a 45° simple offset.

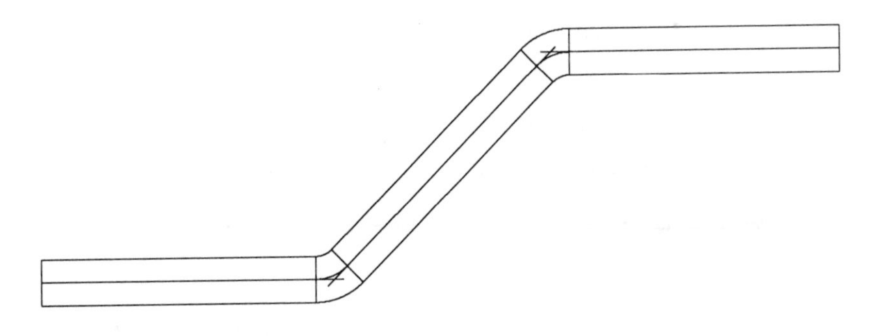

Notice where the offset triangle is in the offset.

The points where the take out lines meet are the vertex of the angles of the offset triangle.

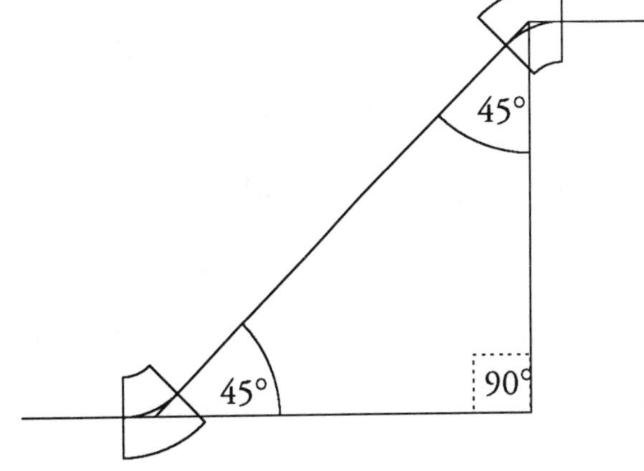

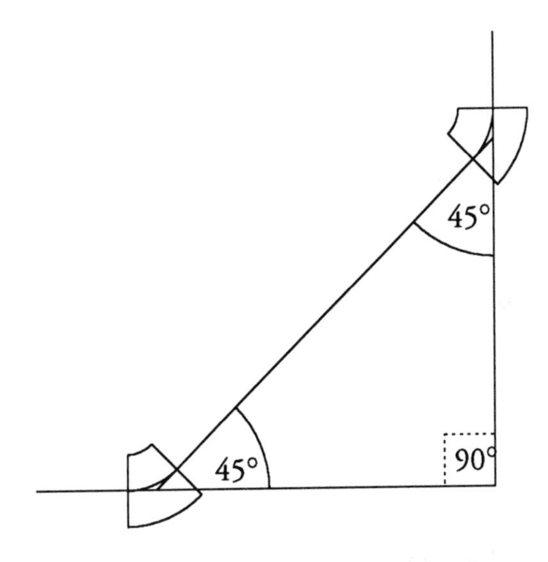

NOTE: You can turn one of the 45° elbows around to make an offset turn. Again, the vertex of the angles of the offset triangles are the points where the take out lines meet.

From the information given in this section, you should now be able to draw, cut out, and put together just about anything you need.

AFTERWORD

This is the end of this book, but not the end of the process for you or us. You have more to learn in your trade and we have more to write about the subject of how math is used to solve problems. Both of our journeys are never ending because each thing learned reveals another thing yet to be learned.

The information in this book comes from centuries of people sharing their knowledge and skills with others. None of it is new. We acknowledge our debt to many sources that we learned from through the years. Some of the information came verbally from crafts people in the field. Some came from schools. Some came from trade and math books.

We learn by sharing. Please add your part by being unselfish and sharing the information and skills you've learned. You have had to work hard to gain these skills, and, if another person is willing to do the same, give him or her a hand.

notes

Glossary

Term	Definition
Acute angle	Any angle between 0° and 90°.
Adjacent side	The side of a right triangle that is next to the reference angle. (Not hypotenuse)
Angle	1. The shape made by two straight lines meeting in a point. 2. The space between those lines. 3. The measurement of that space in degrees or radians.
Angle of turn	The angle created when a line of work is turned. The angle of turn is formed between where the original line would have gone (if it had not turned) and the line going in the new direction.
Arc function (also called inverse function)	The function used to find the degrees of an angle when the ratio of the sides of the angle is known. The six arc functions are: arc sine, arc cosine, arc tangent, arc cosecant, arc secant, and arc cotangent.
Arc	A curved line whose points are equal distances from a single point.
Arc length	The length of a curved line.
Area	The measurement of a surface, expressed in square units.
Bisecting line	A line which divides an object or shape into two equal parts.
Central angle	An angle whose vertex is located at the center of a circle.
Center length	The distance of the straight line or piece of material between the arcs of an offset.
Center Mark Back	The distance used in marking and cutting a miter from the center line.
Chord	A straight line from one point on a circle to another point on the circle. The longest chord of a circle is the diameter.
Circle	A curved line in a plane that encloses a space. Every point on the curved line is exactly the same distance from the center point.
Circumference	The distance around a circle.
Common side of two triangles	A line that is shared by two triangles. It can be the hypotenuse of one triangle and the leg of the other.
Compass	A tool used to draw circles and curved lines.
Complementary angles	Two angles whose sum equals 90°. The two lesser angles of a right triangle are always complementary.

Conversion	For fractions: The process of changing the numerator and the denominator of a fraction to make a new fraction of equal value. Example: $\frac{1}{2}$ to $\frac{2}{4}$ For units of measurement: The process of changing a unit of measurement to a different unit of measurement while keeping the value the same. Example: 2 feet to inches, 2 feet = 24 inches.
Cosecant	One of the functions of an angle. It is found by dividing the length of the hypotenuse of a right triangle by the length of the opposite side of a reference angle.
Cosine	One of the functions of an angle. It is found by dividing the length of the adjacent side of a reference angle by the length of the hypotenuse.
Cotangent	One of the functions of an angle. It is found by dividing the length of the adjacent side of a reference angle by the length of the opposite side of a reference angle.
Decimal	Shortened version of decimal fraction.
Decimal fraction	A fraction in which the denominator is 10 or a power of 10 (such as 100, 1000, 10,000, etc.). The denominator is seldom used. Instead, a dot called a decimal point is placed in front of the numerator. The denominator can be inferred by the number of decimal places behind the decimal point. If there is one number to the right of the decimal point, the denominator is 10. If there are two numbers to the right of the decimal point, the denominator is 100. There will be the same number of zeroes in the denominator as there are decimal places.
Degree	A unit of measurement for angles.(Also see Radians)
Denominator	The bottom number in a fraction. It identifies the number of portions into which the whole has been divided. In the fraction $\frac{9}{16}$ ", the whole inch has been divided into 16 parts, and the measurement is equal to 9 of those parts.
Diameter	A straight line from one side of a circle to the other side that passes through the center. Also the measurement of that line.
Division bar of a fraction	The line between the numerator and the denominator of a fraction.
Equilateral triangle	A triangle with equal sides and equal angles.
Formula	An equation that follows a rule. Formulas contain variables that when replaced by numbers, allow unknown quantities to be found. An example of a formula is the Pythagorean Theorem, $a^2 + b^2 = c^2$.

Fraction	A number that expresses a portion of a whole. The denominator of a fraction represents the number of the portions the whole has been divided into, and the numerator expresses the number of the portions measured. The fraction $\frac{1}{4}$ could be stated as 1 out of 4 parts of the whole.
Fractions, ruler	The portions into which the English units of measurement such as inches can be divided—for example: halves, fourths, eighths, sixteenths, and thirty-seconds.
Functions of an angle	The name of a particular ratio of the sides of a right triangle. The name of the function will depend on which side is divided into which side. There are six possible ways to divide the sides of a right triangle into each other. There are six functions for each angle. The names are: sine, cosine, tangent, cosecant, secant, and cotangent.
Functions table	A table of the functions of angles. This table allows one to find either the function of an angle or the degree of an angle.
Hypotenuse	The longest side of a right triangle. It is always located directly across from the right angle.
Improper fraction	A fraction which is equal to one or more than one.
Inside arc	The shortest arc of a turn.
Inverse	A term which means turned upside down. For example, this ratio, $\frac{\text{top}}{\text{bottom}}$ is expressed inversely as $\frac{\text{bottom}}{\text{top}}$.
Isosceles right triangle	A 45° right triangle. There is only one right triangle that has two equal legs – the 45° right triangle.
Isosceles triangle	A triangle that has two equal legs and two equal angles.
Knowns	Variables in formulas that can be replaced by the actual measurements.
Legs of a right triangle	The sides of a right triangle other than the hypotenuse.
Level	Even, flat, not having any part higher or lower than another part.
Lowest term fraction	A fraction which can not be reduced to a lower term.
Measuring fractions	The fractions used with English units of measurement when measuring distance. *See Fractions, ruler.*
Metric measurement	Measurements which use metric units of measurement.
Minute	A division of a degree. There are 60 minutes in each degree.
Miter	To cut material at an angle.
Mixed number	A whole number and a fraction.
Needs	The unknowns of a formula, or the measurements needed to do a calculation.
Numerator	The top number of a fraction. It indicates the number of portions of a whole.
Obtuse angle	An angle between 90° and 180°.

Offset	To change the direction of a line by making a turn *and* correcting back to the original direction by making another turn. Also, offset is the term used to identify such a turn.
Offset box	An imaginary box which is used when drawing thumbnail sketches to calculate offsets.
Opposite side	The side of a right triangle that is opposite the reference angle.
Outside arc	The longest arc of a turn.
Parallel lines	Lines that are in the same plane and always the same distance apart.
Perpendicular	A 90° angle between two lines. It is also called a right angle.
Pi(π)	The ratio of the circumference of a circle to its diameter. It is the whole number 3 and a decimal fraction that has endless decimal places. We usually round it off to four decimal places (3.1416) for our work.
Plane	A flat, level or even surface.
Plumb	Exactly vertical. A line that is perpendicular to ground level is plumb.
Point of tangency	The point where a circle and a tangent line touch.
Pythagorean theorem	The formula $a^2 + b^2 = c^2$, which signifies that the square of the length of the hypotenuse of a right triangle equals the sum of the squares of the length of the other two sides.
Radians	A measurement of an angle or an arc length based on the length of the radius. One radian is an arc equal in length to the radius. One radian = $\frac{180°}{\pi} = 57.295°$
Radii	More than one radius.
Radius	A straight line from the center of a circle to a point on the circle. Also the measurement of that line.
Radius of an arc (or elbow/bend)	The distance from the vertex of an arc to the center line of the elbow or bend.
Ratio	A comparison of one value to another value by division.
Ratios of the sides	The division of one side of a right triangle by another side. The ratios of the sides are directly related to the number of degrees in the reference angle.
Rectangle	Any four-sided plane with four right angles.
Reference angle	The angle being referred to in any given problem.
Regular polygon	A polygon which has equal angles and equal sides and can be inscribed in a circle.
Right angle	A 90° angle.
Right triangle	A triangle with one angle equal to 90° and two angles whose sum is 90°.
Rolling offset	A simple offset that is rolled to one side.

Rounding off a number Simplifying a number by slightly raising or lowering its value. Accuracy in work is restricted by the tools used in that work. Calculations can not be any closer than what the tools allow, so numbers are rounded off to a point useful for that tool. Needed accuracy will vary among trades or jobs.

Secant One of the functions of an angle. It is found by dividing the length of the hypotenuse of a right triangle by the length of the adjacent side of a reference angle.

Sector A part of a circle which includes two radii connected by the arc.

Second A division of the minutes of a degree. There are 60 seconds in each minute of a degree.

Simple offset A procedure by which a center line is moved up, down, or over to reach a new path going in the same direction. A simple offset uses two same angle turns.

Sine One of the functions of an angle. It is found by dividing the length of the opposite side of a reference angle by the length of the hypotenuse of the right triangle.

Square root A factor of a number which when squared produces that number. A square root multiplied by itself equals the number under the square root symbol. For instances, $\sqrt{9}$ symbolizes the square root of 9, which is 3.

Square The product of a number multiplied by itself.

Straight angle An angle of 180° (in other words, a straight line).

Take out The distance that a fitting extends over the center line of a hypotenuse of an offset triangle.

Take out formula A formula for finding the take out of an elbow. The take out formula is: Take out = Tan $\frac{\theta}{2}$ x radius of the elbow.

Tangent One of the functions of an angle. It is found by dividing the length of the opposite side of a reference angle by the length of the adjacent side of a reference angle.

Tangent line A straight line which touches just one point on a circle.

Theta - θ A letter of the Greek alphabet that is used as a variable for unknown angles or when referring to any angle.

Triangle An enclosed geometric form with three straight sides. The angles of a triangle always equal 180°.

Triangle of roll A triangle in a rolling offset box that contains the angle of roll. It is usually the first triangle calculated.

Turn To change direction.

Unit circle A circle with a radius of 1 unit.

Units of measurement Terms which describe fixed standard measurements.

Variables Letters of the alphabet used as symbols to represent numbers that change or are to be determined.

Vertex The point at which two straight lines come together to form the angle.

Volume The amount of space occupied in three dimensions, expressed in cubic units.

Whole numbers Numbers which contain no fractions.

Zero degree angle An angle with no space between the two lines. In other words, the two lines are occupying the same space. The difference between a straight angle and a zero degree angle is that the vertex is in the middle of the line for a straight angle but at the end of the line for a zero degree angle.

Appendix

Fractions Of An Inch Chart

32nds	16ths	8ths	4ths	Halves	Decimal
$\frac{0}{32}$	$\frac{0}{16}$	$\frac{0}{8}$	$\frac{0}{4}$	$\frac{0}{2}$	0.0000
$\frac{1}{32}$					0.03125
$\frac{2}{32}$	$\frac{1}{16}$				0.0625
$\frac{3}{32}$					0.09375
$\frac{4}{32}$	$\frac{2}{16}$	$\frac{1}{8}$			0.125
$\frac{5}{32}$					0.15625
$\frac{6}{32}$	$\frac{3}{16}$				0.1875
$\frac{7}{32}$					0.21875
$\frac{8}{32}$	$\frac{4}{16}$	$\frac{2}{8}$	$\frac{1}{4}$		0.25
$\frac{9}{32}$					0.28125
$\frac{10}{32}$	$\frac{5}{16}$				0.3125
$\frac{11}{32}$					0.34375
$\frac{12}{32}$	$\frac{6}{16}$	$\frac{3}{8}$			0.375
$\frac{13}{32}$					0.40625
$\frac{14}{32}$	$\frac{7}{16}$				0.4375
$\frac{15}{32}$					0.46875
$\frac{16}{32}$	$\frac{8}{16}$	$\frac{4}{8}$	$\frac{2}{4}$	$\frac{1}{2}$	0.5
$\frac{17}{32}$					0.53125
$\frac{18}{32}$	$\frac{9}{16}$				0.5625
$\frac{19}{32}$					0.59375
$\frac{20}{32}$	$\frac{10}{16}$	$\frac{5}{8}$			0.625
$\frac{21}{32}$					0.65625
$\frac{22}{32}$	$\frac{11}{16}$				0.6875
$\frac{23}{32}$					0.71875
$\frac{24}{32}$	$\frac{12}{16}$	$\frac{6}{8}$	$\frac{3}{4}$		0.75
$\frac{25}{32}$					0.78125
$\frac{26}{32}$	$\frac{13}{16}$				0.8125
$\frac{27}{32}$					0.84375
$\frac{28}{32}$	$\frac{14}{16}$	$\frac{7}{8}$			0.875
$\frac{29}{32}$					0.90625
$\frac{30}{32}$	$\frac{15}{16}$				0.9375
$\frac{31}{32}$					0.96875
$\frac{32}{32}$	$\frac{16}{16}$	$\frac{8}{8}$	$\frac{4}{4}$	$\frac{2}{2}$	1

Conversion Chart For Decimals Of A Foot

1/32"	0.0026'	1 1/32"	0.0859'	2 1/32"	0.1693'	3 1/32"	0.2526'	4 1/32"	0.3359'	5 1/32"	0.4193'
1/16"	0.0052'	1 1/16"	0.0885'	2 1/16"	0.1719'	3 1/16"	0.2552'	4 1/16"	0.3385'	5 1/16"	0.4219'
3/32"	0.0078'	1 3/32"	0.0911'	2 3/32"	0.1745'	3 3/32"	0.2578'	4 3/32"	0.3411'	5 3/32"	0.4245'
1/8"	0.0104'	1 1/8"	0.0938'	2 1/8"	0.1771'	3 1/8"	0.2604'	4 1/8"	0.3438'	5 1/8"	0.4271'
5/32"	0.0130'	1 5/32"	0.0964'	2 5/32"	0.1797'	3 5/32"	0.2630'	4 5/32"	0.3464'	5 5/32"	0.4297'
3/16"	0.0156'	1 3/16"	0.0990'	2 3/16"	0.1823'	3 3/16"	0.2656'	4 3/16"	0.3490'	5 3/16"	0.4323'
7/32"	0.0182'	1 7/32"	0.1016'	2 7/32"	0.1849'	3 7/32"	0.2682'	4 7/32"	0.3516'	5 7/32"	0.4349'
1/4"	0.0208'	1 1/4"	0.1042'	2 1/4"	0.1875'	3 1/4"	0.2708'	4 1/4"	0.3542'	5 1/4"	0.4375'
9/32"	0.0234'	1 9/32"	0.1068'	2 9/32"	0.1901'	3 9/32"	0.2734'	4 9/32"	0.3568'	5 9/32"	0.4401'
5/16"	0.0260'	1 5/16"	0.1094'	2 5/16"	0.1927'	3 5/16"	0.2760'	4 5/16"	0.3594'	5 5/16"	0.4427'
11/32"	0.0286'	1 11/32"	0.1120'	2 11/32"	0.1953'	3 11/32"	0.2786'	4 11/32"	0.3620'	5 11/32"	0.4453'
3/8"	0.0313'	1 3/8"	0.1146'	2 3/8"	0.1979'	3 3/8"	0.2813'	4 3/8"	0.3646'	5 3/8"	0.4479'
13/32"	0.0339'	1 13/32"	0.1172'	2 13/32"	0.2005'	3 13/32"	0.2839'	4 13/32"	0.3672'	5 13/32"	0.4505'
7/16"	0.0365'	1 7/16"	0.1198'	2 7/16"	0.2031'	3 7/16"	0.2865'	4 7/16"	0.3698'	5 7/16"	0.4531'
15/32"	0.0391'	1 15/32"	0.1224'	2 15/32"	0.2057'	3 15/32"	0.2891'	4 15/32"	0.3724'	5 15/32"	0.4557'
1/2"	0.0417'	1 1/2"	0.1250'	2 1/2"	0.2083'	3 1/2"	0.2917'	4 1/2"	0.3750'	5 1/2"	0.4583'
17/32"	0.0443'	1 17/32"	0.1276'	2 17/32"	0.2109'	3 17/32"	0.2943'	4 17/32"	0.3776'	5 17/32"	0.4609'
9/16"	0.0469'	1 9/16"	0.1302'	2 9/16"	0.2135'	3 9/16"	0.2969'	4 9/16"	0.3802'	5 9/16"	0.4635'
19/32"	0.0495'	1 19/32"	0.1328'	2 19/32"	0.2161'	3 19/32"	0.2995'	4 19/32"	0.3828'	5 19/32"	0.4661'
5/8"	0.0521'	1 5/8"	0.1354'	2 5/8"	0.2188'	3 5/8"	0.3021'	4 5/8"	0.3854'	5 5/8"	0.4688'
21/32"	0.0547'	1 21/32"	0.1380'	2 21/32"	0.2214'	3 21/32"	0.3047'	4 21/32"	0.3880'	5 21/32"	0.4714'
11/16"	0.0573'	1 11/16"	0.1406'	2 11/16"	0.2240'	3 11/16"	0.3073'	4 11/16"	0.3906'	5 11/16"	0.4740'
23/32"	0.0599'	1 23/32"	0.1432'	2 23/32"	0.2266'	3 23/32"	0.3099'	4 23/32"	0.3932'	5 23/32"	0.4766'
3/4"	0.0625'	1 3/4"	0.1458'	2 3/4"	0.2292'	3 3/4"	0.3125'	4 3/4"	0.3958'	5 3/4"	0.4792'
25/32"	0.0651'	1 25/32"	0.1484'	2 25/32"	0.2318'	3 25/32"	0.3151'	4 25/32"	0.3984'	5 25/32"	0.4818'
13/16"	0.0677'	1 13/16"	0.1510'	2 13/16"	0.2344'	3 13/16"	0.3177'	4 13/16"	0.4010'	5 13/16"	0.4844'
27/32"	0.0703'	1 27/32"	0.1536'	2 27/32"	0.2370'	3 27/32"	0.3203'	4 27/32"	0.4036'	5 27/32"	0.4870'
7/8"	0.0729'	1 7/8"	0.1563'	2 7/8"	0.2396'	3 7/8"	0.3229'	4 7/8"	0.4063'	5 7/8"	0.4896'
29/32"	0.0755'	1 29/32"	0.1589'	2 29/32"	0.2422'	3 29/32"	0.3255'	4 29/32"	0.4089'	5 29/32"	0.4922'
15/16"	0.0781'	1 15/16"	0.1615'	2 15/16"	0.2448'	3 15/16"	0.3281'	4 15/16"	0.4115'	5 15/16"	0.4948'
31/32"	0.0807'	1 31/32"	0.1641'	2 31/32"	0.2474'	3 31/32"	0.3307'	4 31/32"	0.4141'	5 31/32"	0.4974'
1"	0.0833'	2"	0.1667'	3"	0.2500'	4"	0.3333'	5"	0.4167'	6"	0.5000'

Conversion Chart For Decimals Of A Foot

6 1/32"	0.5026'	7 1/32"	0.5859'	8 1/32"	0.6693'	9 1/32"	0.7526'	10 1/32"	0.8359	11 1/32"	0.9193'
6 1/16"	0.5052'	7 1/16"	0.5885'	8 1/16"	0.6719'	9 1/16"	0.7552'	10 1/16"	0.8385'	11 1/16"	0.9219'
6 3/32"	0.5078'	7 3/32"	0.5911'	8 3/32"	0.6745'	9 3/32"	0.7578'	10 3/32"	0.8411'	11 3/32"	0.9245'
6 1/8"	0.5104'	7 1/8"	0.5938'	8 1/8"	0.6771'	9 1/8"	0.7604'	10 1/8"	0.8438'	11 1/8"	0.9271'
6 5/32"	0.5130'	7 5/32"	0.5964'	8 5/32"	0.6797'	9 5/32"	0.7630'	10 5/32"	0.8464'	11 5/32"	0.9297'
6 3/16"	0.5156'	7 3/16"	0.5990'	8 3/16"	0.6823'	9 3/16"	0.7656'	10 3/16"	0.8490'	11 3/16"	0.9323'
6 7/32"	0.5182'	7 7/32"	0.6016'	8 7/32"	0.6849'	9 7/32"	0.7682'	10 7/32"	0.8516'	11 7/32"	0.9349'
6 1/4"	0.5208'	7 1/4"	0.6042'	8 1/4"	0.6875'	9 1/4"	0.7708'	10 1/4"	0.8542'	11 1/4"	0.9375'
6 9/32"	0.5234'	7 9/32"	0.6068'	8 9/32"	0.6901'	9 9/32"	0.7734'	10 9/32"	0.8568'	11 9/32"	0.9401'
6 5/16"	0.5260'	7 5/16"	0.6094'	8 5/16"	0.6927'	9 5/16"	0.7760'	10 5/16"	0.8594'	11 5/16"	0.9427'
6 11/32"	0.5286'	7 11/32"	0.6120'	8 11/32"	0.6953'	9 11/32"	0.7786'	10 11/32"	0.8620'	11 11/32"	0.9453'
6 3/8"	0.5313'	7 3/8"	0.6146'	8 3/8"	0.6979'	9 3/8"	0.7813'	10 3/8"	0.8646'	11 3/8"	0.9479'
6 13/32"	0.5339'	7 13/32"	0.6172'	8 13/32"	0.7005'	9 13/32"	0.7839'	10 13/32"	0.8672'	11 13/32"	0.9505'
6 7/16"	0.5365'	7 7/16"	0.6198'	8 7/16"	0.7031'	9 7/16"	0.7865'	10 7/16"	0.8698'	11 7/16"	0.9531'
6 15/32"	0.5391'	7 15/32"	0.6224'	8 15/32"	0.7057'	9 15/32"	0.7891'	10 15/32"	0.8724'	11 15/32"	0.9557'
6 1/2"	0.5417'	7 1/2"	0.6250'	8 1/2"	0.7083'	9 1/2"	0.7917'	10 1/2"	0.8750'	11 1/2"	0.9583'
6 17/32"	0.5443'	7 17/32"	0.6276'	8 17/32"	0.7109'	9 17/32"	0.7943'	10 17/32"	0.8776'	11 17/32"	0.9609'
6 9/16"	0.5469'	7 9/16"	0.6302'	8 9/16"	0.7135'	9 9/16"	0.7969'	10 9/16"	0.8802'	11 9/16"	0.9635'
6 19/32"	0.5495'	7 19/32"	0.6328'	8 19/32"	0.7161'	9 19/32"	0.7995'	10 19/32"	0.8828'	11 19/32"	0.9661'
6 5/8"	0.5521'	7 5/8"	0.6354'	8 5/8"	0.7188'	9 5/8"	0.8021'	10 5/8"	0.8854'	11 5/8"	0.9688'
6 21/32"	0.5547'	7 21/32"	0.6380'	8 21/32"	0.7214'	9 21/32"	0.8047'	10 21/32"	0.8880'	11 21/32"	0.9714'
6 11/16"	0.5573'	7 11/16"	0.6406'	8 11/16"	0.7240'	9 11/16"	0.8073'	10 11/16"	0.8906'	11 11/16"	0.9740'
6 23/32"	0.5599'	7 23/32"	0.6432'	8 23/32"	0.7266'	9 23/32"	0.8099'	10 23/32"	0.8932'	11 23/32"	0.9766'
6 3/4"	0.5625'	7 3/4"	0.6458'	8 3/4"	0.7292'	9 3/4"	0.8125'	10 3/4"	0.8958'	11 3/4"	0.9792'
6 25/32"	0.5651'	7 25/32"	0.6484'	8 25/32"	0.7318'	9 25/32"	0.8151'	10 25/32"	0.8984'	11 25/32"	0.9818'
6 13/16"	0.5677'	7 13/16"	0.6510'	8 13/16"	0.7344'	9 13/16"	0.8177'	10 13/16"	0.9010'	11 13/16"	0.9844'
6 27/32"	0.5703'	7 27/32"	0.6536'	8 27/32"	0.7370'	9 27/32"	0.8203'	10 27/32"	0.9036'	11 27/32"	0.9870'
6 7/8"	0.5729'	7 7/8"	0.6563'	8 7/8"	0.7396'	9 7/8"	0.8229'	10 7/8"	0.9063'	11 7/8"	0.9896'
6 29/32"	0.5755'	7 29/32"	0.6589'	8 29/32"	0.7422'	9 29/32"	0.8255'	10 29/32"	0.9089'	11 29/32"	0.9922'
6 15/16"	0.5781'	7 15/16"	0.6615'	8 15/16"	0.7448'	9 15/16"	0.8281'	10 15/16"	0.9115'	11 15/16"	0.9948'
6 31/32"	0.5807'	7 31/32"	0.6641'	8 31/32"	0.7474'	9 31/32"	0.8307'	10 31/32"	0.9141'	11 31/32"	0.9974'
7"	0.5833'	8"	0.6667'	9"	0.7500'	10"	0.8333'	11"	0.9167'	12"	1'

Functions Table

Deg ↓	Radian↓	Sin θ↓	Cos θ↓	Tan θ↓	Cot θ↓	Sec θ ↓	Csc θ↓		
0°	0.0000	0.0000	1.0000	0.0000	-	1.0000	-	1.5708	90°
0.5°	0.0087	0.0087	1.0000	0.0087	114.589	1.0000	114.593	1.5621	89.5°
1°	0.0175	0.0175	0.9998	0.0175	57.2900	1.0002	57.2987	1.5533	89°
1.5°	0.0262	0.0262	0.9997	0.0262	38.1885	1.0003	38.2016	1.5446	88.5°
2°	0.0349	0.0349	0.9994	0.0349	28.6363	1.0006	28.6537	1.5359	88°
2.5°	0.0436	0.0436	0.9990	0.0437	22.9038	1.0010	22.9256	1.5272	87.5°
3°	0.0524	0.0523	0.9986	0.0524	19.0811	1.0014	19.1073	1.5184	87°
3.5°	0.0611	0.0610	0.9981	0.0612	16.3499	1.0019	16.3804	1.5097	86.5°
4°	0.0698	0.0698	0.9976	0.0699	14.3007	1.0024	14.3356	1.5010	86°
4.5°	0.0785	0.0785	0.9969	0.0787	12.7062	1.0031	12.7455	1.4923	85.5°
5°	0.0873	0.0872	0.9962	0.0875	11.4301	1.0038	11.4737	1.4835	85°
5.5°	0.0960	0.0958	0.9954	0.0963	10.3854	1.0046	10.4334	1.4748	84.5°
6°	0.1047	0.1045	0.9945	0.1051	9.5144	1.0055	9.5668	1.4661	84°
6.5°	0.1134	0.1132	0.9936	0.1139	8.7769	1.0065	8.8337	1.4573	83.5°
7°	0.1222	0.1219	0.9925	0.1228	8.1443	1.0075	8.2055	1.4486	83°
7.5°	0.1309	0.1305	0.9914	0.1317	7.5958	1.0086	7.6613	1.4399	82.5°
8°	0.1396	0.1392	0.9903	0.1405	7.1154	1.0098	7.1853	1.4312	82°
8.5°	0.1484	0.1478	0.9890	0.1495	6.6912	1.0111	6.7655	1.4224	81.5°
9°	0.1571	0.1564	0.9877	0.1584	6.3138	1.0125	6.3925	1.4137	81°
9.5°	0.1658	0.1650	0.9863	0.1673	5.9758	1.0139	6.0589	1.4050	80.5°
10°	0.1745	0.1736	0.9848	0.1763	5.6713	1.0154	5.7588	1.3963	80°
10.5°	0.1833	0.1822	0.9833	0.1853	5.3955	1.0170	5.4874	1.3875	79.5°
11°	0.1920	0.1908	0.9816	0.1944	5.1446	1.0187	5.2408	1.3788	79°
11.5°	0.2007	0.1994	0.9799	0.2035	4.9152	1.0205	5.0159	1.3701	78.5°
12°	0.2094	0.2079	0.9781	0.2126	4.7046	1.0223	4.8097	1.3614	78°
12.5°	0.2182	0.2164	0.9763	0.2217	4.5107	1.0243	4.6202	1.3526	77.5°
13°	0.2269	0.2250	0.9744	0.2309	4.3315	1.0263	4.4454	1.3439	77°
13.5°	0.2356	0.2334	0.9724	0.2401	4.1653	1.0284	4.2837	1.3352	76.5°
14°	0.2443	0.2419	0.9703	0.2493	4.0108	1.0306	4.1336	1.3265	76°
14.5°	0.2531	0.2504	0.9681	0.2586	3.8667	1.0329	3.9939	1.3177	75.5°
15°	0.2618	0.2588	0.9659	0.2679	3.7321	1.0353	3.8637	1.3090	75°
15.5°	0.2705	0.2672	0.9636	0.2773	3.6059	1.0377	3.7420	1.3003	74.5°
16°	0.2793	0.2756	0.9613	0.2867	3.4874	1.0403	3.6280	1.2915	74°
16.5°	0.2880	0.2840	0.9588	0.2962	3.3759	1.0429	3.5209	1.2828	73.5°
17°	0.2967	0.2924	0.9563	0.3057	3.2709	1.0457	3.4203	1.2741	73°
17.5°	0.3054	0.3007	0.9537	0.3153	3.1716	1.0485	3.3255	1.2654	72.5°
18°	0.3142	0.3090	0.9511	0.3249	3.0777	1.0515	3.2361	1.2566	72°
18.5°	0.3229	0.3173	0.9483	0.3346	2.9887	1.0545	3.1515	1.2479	71.5°
19°	0.3316	0.3256	0.9455	0.3443	2.9042	1.0576	3.0716	1.2392	71°
19.5°	0.3403	0.3338	0.9426	0.3541	2.8239	1.0608	2.9957	1.2305	70.5°
20°	0.3491	0.3420	0.9397	0.3640	2.7475	1.0642	2.9238	1.2217	70°
20.5°	0.3578	0.3502	0.9367	0.3739	2.6746	1.0676	2.8555	1.2130	69.5°
21°	0.3665	0.3584	0.9336	0.3839	2.6051	1.0711	2.7904	1.2043	69°
21.5°	0.3752	0.3665	0.9304	0.3939	2.5386	1.0748	2.7285	1.1956	68.5°
22°	0.3840	0.3746	0.9272	0.4040	2.4751	1.0785	2.6695	1.1868	68°
		Cos θ ↑	Sin θ ↑	Cot θ ↑	Tan θ ↑	Csc θ ↑	Sec θ ↑	Radian↑	↑Deg

Functions Table

Deg↓	Radian↓	Sin θ ↓	Cos θ ↓	Tan θ ↓	Cot θ ↓	Sec θ ↓	Csc θ ↓		
22.5°	0.3927	0.3827	0.9239	0.4142	2.4142	1.0824	2.6131	1.1781	67.5°
23°	0.4014	0.3907	0.9205	0.4245	2.3559	1.0864	2.5593	1.1694	67°
23.5°	0.4102	0.3987	0.9171	0.4348	2.2998	1.0904	2.5078	1.1606	66.5°
24°	0.4189	0.4067	0.9135	0.4452	2.2460	1.0946	2.4586	1.1519	66°
24.5°	0.4276	0.4147	0.9100	0.4557	2.1943	1.0989	2.4114	1.1432	65.5°
25°	0.4363	0.4226	0.9063	0.4663	2.1445	1.1034	2.3662	1.1345	65°
25.5°	0.4451	0.4305	0.9026	0.4770	2.0965	1.1079	2.3228	1.1257	64.5°
26°	0.4538	0.4384	0.8988	0.4877	2.0503	1.1126	2.2812	1.1170	64°
26.5°	0.4625	0.4462	0.8949	0.4986	2.0057	1.1174	2.2412	1.1083	63.5°
27°	0.4712	0.4540	0.8910	0.5095	1.9626	1.1223	2.2027	1.0996	63°
27.5°	0.4800	0.4617	0.8870	0.5206	1.9210	1.1274	2.1657	1.0908	62.5°
28°	0.4887	0.4695	0.8829	0.5317	1.8807	1.1326	2.1301	1.0821	62°
28.5°	0.4974	0.4772	0.8788	0.5430	1.8418	1.1379	2.0957	1.0734	61.5°
29°	0.5061	0.4848	0.8746	0.5543	1.8040	1.1434	2.0627	1.0647	61°
29.5°	0.5149	0.4924	0.8704	0.5658	1.7675	1.1490	2.0308	1.0559	60.5°
30°	0.5236	0.5000	0.8660	0.5774	1.7321	1.1547	2.0000	1.0472	60°
30.5°	0.5323	0.5075	0.8616	0.5890	1.6977	1.1606	1.9703	1.0385	59.5°
31°	0.5411	0.5150	0.8572	0.6009	1.6643	1.1666	1.9416	1.0297	59°
31.5°	0.5498	0.5225	0.8526	0.6128	1.6319	1.1728	1.9139	1.0210	58.5°
32°	0.5585	0.5299	0.8480	0.6249	1.6003	1.1792	1.8871	1.0123	58°
32.5°	0.5672	0.5373	0.8434	0.6371	1.5697	1.1857	1.8612	1.0036	57.5°
33°	0.5760	0.5446	0.8387	0.6494	1.5399	1.1924	1.8361	0.9948	57°
33.5°	0.5847	0.5519	0.8339	0.6619	1.5108	1.1992	1.8118	0.9861	56.5°
34°	0.5934	0.5592	0.8290	0.6745	1.4826	1.2062	1.7883	0.9774	56°
34.5°	0.6021	0.5664	0.8241	0.6873	1.4550	1.2134	1.7655	0.9687	55.5°
35°	0.6109	0.5736	0.8192	0.7002	1.4281	1.2208	1.7434	0.9599	55°
35.5°	0.6196	0.5807	0.8141	0.7133	1.4019	1.2283	1.7221	0.9512	54.5°
36°	0.6283	0.5878	0.8090	0.7265	1.3764	1.2361	1.7013	0.9425	54°
36.5°	0.6370	0.5948	0.8039	0.7400	1.3514	1.2440	1.6812	0.9338	53.5°
37°	0.6458	0.6018	0.7986	0.7536	1.3270	1.2521	1.6616	0.9250	53°
37.5°	0.6545	0.6088	0.7934	0.7673	1.3032	1.2605	1.6427	0.9163	52.5°
38°	0.6632	0.6157	0.7880	0.7813	1.2799	1.2690	1.6243	0.9076	52°
38.5°	0.6720	0.6225	0.7826	0.7954	1.2572	1.2778	1.6064	0.8988	51.5°
39°	0.6807	0.6293	0.7771	0.8098	1.2349	1.2868	1.5890	0.8901	51°
39.5°	0.6894	0.6361	0.7716	0.8243	1.2131	1.2960	1.5721	0.8814	50.5°
40°	0.6981	0.6428	0.7660	0.8391	1.1918	1.3054	1.5557	0.8727	50°
40.5°	0.7069	0.6494	0.7604	0.8541	1.1708	1.3151	1.5398	0.8639	49.5°
41°	0.7156	0.6561	0.7547	0.8693	1.1504	1.3250	1.5243	0.8552	49°
41.5°	0.7243	0.6626	0.7490	0.8847	1.1303	1.3352	1.5092	0.8465	48.5°
42°	0.7330	0.6691	0.7431	0.9004	1.1106	1.3456	1.4945	0.8378	48°
42.5°	0.7418	0.6756	0.7373	0.9163	1.0913	1.3563	1.4802	0.8290	47.5°
43°	0.7505	0.6820	0.7314	0.9325	1.0724	1.3673	1.4663	0.8203	47°
43.5°	0.7592	0.6884	0.7254	0.9490	1.0538	1.3786	1.4527	0.8116	46.5°
44°	0.7679	0.6947	0.7193	0.9657	1.0355	1.3902	1.4396	0.8029	46°
44.5°	0.7767	0.7009	0.7133	0.9827	1.0176	1.4020	1.4267	0.7941	45.5°
45°	0.7854	0.7071	0.7071	1.0000	1.0000	1.4142	1.4142	0.7854	45°
		Cos θ ↑	Sin θ ↑	Cot θ ↑	Tan θ ↑	Csc θ ↑	Sec θ ↑	Radian ↑	↑Deg

notes

Answers to Practices

We have verified these answers many times to ensure their accuracy; However, mistakes do get past us. If you find a difference in one of your answers and a stated answer, first, rework the problem. If both answers still disagree and your answers for the other problems of that practice agree, then the answer in the back of the book may be incorrect. If most of your answers for a practice disagree with the stated answers, you are probably doing something wrong. Go back to that particular section and reread it, work the examples, and then rework the practice. If you still have difficulty, ask for help. Asking for help is a sign of strength, not weakness. Since this book is written using math terms, you can ask a math teacher, an engineer or a coworker.

Practice 1: (Page 7)

(1) $\frac{1}{8}$ (2) $\frac{1}{2}$ (3) $\frac{3}{8}$ (4) $\frac{5}{8}$ (5) $\frac{3}{4}$ (6) $\frac{1}{2}$ (7) $\frac{7}{8}$ (8) $\frac{1}{4}$

Practice 2: (Page 8)

(1) $\frac{6}{8}$ (2) $\frac{60}{64}$ (3) $\frac{2}{32}$ (4) $\frac{6}{16}$ (5) $\frac{8}{32}$ (6) $\frac{4}{16}$ (7) $\frac{14}{16}$ (8) $\frac{6}{32}$

Practice 3: (Page 9)

(1) 4 (2) 8 (3) 16 (4) 32 (5) 64 (6) 16 (7) 64 (8) 64

Practice 4: (Page 10)

(1) 1 (2) $1\frac{1}{4}$ (3) $\frac{1}{8}$ (4) $\frac{1}{16}$ (5) $\frac{43}{64}$ (6) $\frac{3}{8}$ (7) $2\frac{5}{16}$ (8) $\frac{37}{64}$

Practice 5: (Page 10)

(1) $\frac{1}{8}$ (2) $\frac{9}{16}$ (3) $\frac{9}{48}$ (4) $\frac{2}{15}$ (5) $\frac{55}{128}$ (6) $\frac{9}{32}$ (7) $\frac{3}{96}$ (8) $\frac{6}{1024}$

Practice 6: (Page 11)

(1) $\frac{18}{16}$ or $1\frac{1}{8}$ (2) $\frac{28}{8}$ or $3\frac{1}{2}$ (3) 1 (4) $\frac{240}{48}$ or 5 (5) $\frac{104}{16}$ or $6\frac{1}{2}$ (6) $\frac{96}{16}$ or 6

Practice 7: (Page 13)

(1) $\frac{2}{2}$ or 1 (2) $\frac{60}{8}$ or 7.5 (3) $\frac{7}{16}$ (4) 9+2=11 (5) $20+7\frac{1}{2}=27\frac{1}{2}$ (6) 128 + 5 = 133

Practice 8: (Page 15)

(1) $\frac{875}{1000}$ (2) $\frac{4}{10}$ (3) $\frac{5}{1000}$ (4) $\frac{3}{100}$ (5) $\frac{1507}{10000}$ (6) $\frac{14}{1000}$

Practice 9: (Page 17)

(1) .375" (2) .6875" (3) .4375" (4) .5" (5) .75" (6) .59375"
(7) .875" (8) .1875" (9) .125" (10) .25" (11) .9375" (12) .0625"

Practice 10: (Page 17)

(1) 19.5" (2) 43.6875" (3) 3.9375" (4) 5.3125"
(5) 17.125" (6) 37.875" (7) 7.03125" (8) 24.375"

Practice 11: (Page 18)

(1) $1\frac{3}{12}' = 1.25'$ (2) $7\frac{1}{12}' = 7.0833'$ (3) $62\frac{10}{12}' = 62.8333'$

(4) $29\frac{2}{12}' = 29.1667'$ (5) $44\frac{4}{12}' = 44.3333'$ (6) $137\frac{5}{12}' = 137.4167'$

Practice 12: (Page 20)

(1) $5\frac{6}{12}' = 5.5'$ (2) $2\frac{9}{12}' = 2.75'$ (3) $3\frac{7.75}{12}' = 3.6458'$

(4) $13\frac{8.1875}{12}' = 13.6823'$ (5) $2\frac{4.125}{12}' = 2.3438'$ (6) $283\frac{.625}{12}' = 283.0521'$

(7) $42\frac{8.4375}{12}' = 42.7031'$ (8) $1\frac{11.8125}{12}' = 1.9844'$ (9) $11\frac{1.25}{12}' = 11.1042'$ (10) $87\frac{6.9375}{12}' = 87.5781'$

Practice 13: (Page 22)

(1) $\frac{9}{16}''$ (2) $\frac{3}{4}''$ (3) $\frac{5}{16}''$ (4) 1" (5) $\frac{13}{16}''$ (6) $\frac{1}{8}''$ (7) $\frac{13}{16}''$ (8) $\frac{5}{16}''$

Practice 14: (Page 22)

(1) $12\frac{3}{8}''$ (2) $4\frac{5}{8}''$ (3) 100" (4) $5\frac{3}{4}''$ (5) $84\frac{3}{16}''$ (6) $12\frac{9}{16}''$ (7) $121\frac{13}{16}''$ (8) $23\frac{7}{8}''$

Practice 15: (Page 24)

(1) 12' 9" (2) 1' $10\frac{13}{16}''$ (3) 3' 4" (4) 155' $10\frac{3}{8}''$ (5) 50' $1\frac{1}{2}''$ (6) 40' $3\frac{1}{2}''$

(7) 33' $3\frac{3}{4}''$ (8) 75' $1\frac{9}{16}''$ (9) 84' $5\frac{5}{16}''$ (10) 9' $6\frac{13}{16}''$ (11) 11' $3\frac{9}{16}''$ (12) 14' $11\frac{15}{16}''$

Practice 16: (Page 36)

(1) 89.1875° (2) 12.25° (3) 33.5° (4) 71.0083°
(5) 42.4147° (6) 38.7069° (7) 29.5083° (8) 0.8303°

Practice 17: (Page 38)

(1) 75° 15' 0" (2) 45° 22' 30" (3) 9° 33' 45" (4) 33° 57' 52.2"
(5) 13° 7' 24.42" (6) 21° 30' 0" (7) 59° 47' 21.12" (8) 65° 11' 0.96"

Practice 18: (Page 44)

(1) 1 (2) 100 (3) 16900 (4) 56.25 (5) 30.5256 (6) 70.1406 (7) 121 (8) d^2

Practice 19: (Page 46)

(1) 1 (2) 4.6904 (3) 7.3824 (4) 4.0927 (5) 20 (6) 11.2694
(7) 8.6328 (8) 5.9896 (9) a (10) b (11) 12 (12) 10

Practice 20: (Page 48)

(1) Yes (2) Yes (3) No (4) Yes (5) Yes
(6) Yes (7) Yes (8) No (9) No (10) Yes

Practice 21: (Page 50)

(1) 10.8167 (2) 11.1803 (3) $39\frac{11}{16}''$ (4) $7\frac{7}{8}''$ (5) $12\frac{3}{4}''$ (6) 8' $5\frac{1}{8}''$

(7) 15'$11\frac{11}{16}''$ (8) $2\frac{7}{8}''$ (9) 129'$11\frac{7}{16}''$ (10) 31.4006 (11) $28\frac{1}{2}''$ (12) 50.1199

Practice 22: (Page 52)

(1) 13.2288 (2) $8\frac{3}{16}''$ (3) 13.4164 (4) 6' $11\frac{5}{8}''$ (5) 47' $5\frac{15}{16}''$ (6) $12\frac{5}{8}''$

(7) 10.6677 (8) 40'$6\frac{3}{8}''$ (9) 15.4919 (10) $1\frac{13}{16}''$ (11) 8'$7\frac{15}{16}''$ (12) 1'$8\frac{5}{8}''$

Practice 23: (Page 55)

	Sine	Cosine	Tangent	Cosecant	Secant	Cotangent
A	0.4223	0.9067	0.4657	2.3681	1.1029	2.1472
B	0.9067	0.4223	2.1472	1.1029	2.3681	0.4657
C	0.7071	0.7071	1.0000	1.4142	1.4142	1.0000
D	0.7071	0.7071	1.0000	1.4142	1.4142	1.0000

Practice 24: (Page 59)

(1)	(2)	(3)	(4)	(5)	(6)	(7)	(8)
22°	45°	30.5°	90°	30°	51°	20°	56.5°

Practice 25: (Page 61)

(1) 40.5°/49.5° (2) 55°/35° (3) 34°/56° (4) 26°/64° (5) 56.5°/33.5°

Practice 26: (Page 66)

(1) 28°/62° (2) 53°/37° (3) 28°/62° (4) 56.5°/33.5°

Practice 27: (Page 67)

(1)	1.0000	(2)	0.5446	(3)	0.9063	(4)	0.8660
(5)	0.5878	(6)	3.7321	(7)	0.9998	(8)	0.5000

Practice 28: (Page 69)

(1)	1.5557	(2)	1.4142	(3)	1.4826	(4)	1.0223
(5)	0.5774	(6)	2.4586	(7)	2.3662	(8)	57.29

Practice 29: (Page 70)

(1) 80° (2) 76° (3) 25° (4) 4.5° (5) 63° (6) 48° (7) 22.5° (8) 78.5°

NOTE: Regarding your final answers for many of the practices below, don't be concerned if your calculations indicate a difference of a hundredth or so from the stated answer. You may have calculated a problem using a constant (1.4142 or .7071) while I derived the final answer using a function in my calculator, or I may have used a constant and you calculated using a function in your calculator. In the field, if you are calculating for machining parts or need a high degree of accuracy, use the function in your calculator, not a constant.

Practice 30: (Page 73)

	Length	*Side*		*Length*	*Side*		*Length*	*Side*		*Length*	*Side*
(1)	4.4135"	hyp	(2)	3.5355'	opp	(3)	4.3676"	adj	(4)	57.29"	adj
	1.8652"	adj		3.5355'	adj		12.7701"	hyp		57.2987"	hyp
(5)	7.5128'	hyp	(6)	2.1006'	adj	(7)	127.7982"	opp	(8)	98.9397'	opp
	4.5213'	opp		3.6623'	hyp		46.5147"	adj		3.4551'	adj

Practice 31: (Page 77)

	(1)	(2)	(3)	(4)	(5)	(6)	(7)	(8)
Equal angle	60°	78.25°	67.5°	54°	45°	52.5°	81°	75°
Base	22"	6.9238"	47.6402"	63.4808"	19.799'	5.4788"	12.8276"	40.3758'

Practice 32: (Page 80)

	(1)	(2)	(3)	(4)	(5)	(6)	(7)	(8)
Leg	7"	12.0208'	23"	22.3181"	63.125"	30.4056'	75.5"	62.8441"
Hypo	9.8995"	17'	32.5269"	31.5625"	89.2722"	43'	106.7731"	88.875"

Practice 33: (Page 82)

	(1)	(2)	(3)	(4)	(5)	(6)
Opp	34'	8.5"	46"	8'	56.25"	39.2813"
Hypo	68'	17"	92"	16'	112.5"	78.5625"

Practice 34: (Page 87)

(1)	9' $5\frac{1}{8}$"	(2)	$14\frac{1}{8}$"	(3)	$34\frac{3}{16}$"	(4)	24' $4\frac{3}{16}$"
(5)	$28\frac{1}{4}$"	(6)	$11\frac{3}{4}$"	(7)	50' $3\frac{3}{16}$"	(8)	87' $11\frac{9}{16}$"

Practice 35: (Page 88)

(1)	$12\frac{9}{16}$"	(2)	94'3"	(3)	$119\frac{3}{8}$"	(4)	772' 10"
(5)	$19\frac{5}{8}$"	(6)	12' $6\frac{13}{16}$"	(7)	$95\frac{13}{16}$"	(8)	314' $1\frac{15}{16}$"

Practice 36: (Page 90)

(1) $\frac{4}{12}$ or $\frac{1}{3}$ (2) $\frac{90}{360}$ or $\frac{1}{4}$ (3) $\frac{12}{12.5664}$ (4) $\frac{12}{360}$ or $\frac{1}{30}$

Practice 37: (Page 92)

(1) 7.8540" (2) 4.7124" (3) 2.6180' (4) 2.6180" (5) 3.9270' (6) 1.1781 (7) 3.7088' 8) 1956.3352"

Practice 38: (Page 92)

(1) 120° (2) 344° (3) 52° (4) 30° (5) 15° (6) 6.5° (7) 36.5° (8) 16°

Practice 39: (Page 95)

(1)	0.7854	(2)	0.1745	(3)	2.0944	(4)	6.1087
(5)	1.3788	(6)	0.0175	(7)	0.2531	(8)	1.7366

Practice 40: (Page 96)

(1)	0.3491	(2)	0.7854	(3)	0.6458	(4)	1.0472
(5)	0.0175	(6)	2.2515	(7)	0.0349	(8)	3.9095

Practice 41: (Page 97)

(1) $8\frac{1}{2}$" (2) $11\frac{5}{16}$" (3) $3\frac{9}{16}$" (4) $6\frac{5}{16}$" (5) $21\frac{5}{8}$" (6) 1' $3\frac{1}{16}$"

(7) 5'$9\frac{1}{8}$" (8) $5\frac{7}{8}$" (9) 117' $0\frac{15}{16}$" (10) $10\frac{1}{4}$" (11) $4\frac{3}{4}$" (12) $174\frac{9}{16}$"

Practice 42: (Page 100)

(1)	19.2°	(2)	4.8°	(3)	14.4°	(4)	3.9°
(5)	97.2°	(6)	12°	(7)	29°	(8)	155.5°

Practice 43: (Page 101)

(1) $34\frac{7}{16}$" (2) 14' $1\frac{11}{12}$" (3) $10\frac{9}{16}$" (4) $3\frac{7}{16}$" (5) 9' $2\frac{7}{16}$" (6) 30'

Practice 44: (Page 114)

(1) $14\frac{1}{8}$" (2) 31' $1\frac{3}{8}$" (3) $10\frac{7}{16}$" (4) $60\frac{13}{16}$"

(5) $29\frac{7}{8}$" (6) $110\frac{5}{16}$" (7) $21\frac{1}{16}$" (8) $80\frac{11}{16}$"

Practice 45: (Page 115)

	Hypo	Adj		Hypo	Adj
(1)	150.5"	130.3368"	(5)	90'	77.9423'
(2)	27.125"	23.4909"	(6)	52.25"	45.2498"
(3)	15"	12.9904"	(7)	8'	6.9282'
(4)	6"	5.1962"	(8)	22.5'	19.4856'

Practice 46: (Page 118)

(1)	19.7231"	30.5°	(2)	6.0179'	69.5°	(3)	51.3128"	57°	(4)	8.2890'	65.5°
(5)	35.7780"	69.5°	(6)	57.3847'	22.5°	(7)	7.6526"	38.5°	(8)	5'	37°

Practice 47: (Page 119)

(1)	59.5°	30.5°	27.7849"	(5)	45°	45°	5.6569'
(2)	49.5°	40.5°	82.9759"	(6)	70°	20°	40.4969"
(3)	45°	45°	16.9706'	(7)	30°	60°	101.9960"
(4)	87.5°	2.5°	1225.0208"	(8)	53°	37°	53.3987"

Practice 48: (Page 122)

	Inside arc	Center Arc	Outside Arc		Inside arc	Center Arc	Outside Arc
(1)	5.7596"	6.9813"	8.2030"	(5)	4.0355"	4.2761"	4.5166"
(2)	4.6469'	5.4978'	6.3486'	(6)	0.6981'	1.0472'	1.3963'
(3)	9.7738"	13.1947"	16.6155"	(7)	37.6991"	56.5487"	75.3982"
(4)	3.4034"	5.4454"	7.4875"	(8)	137.4447'	157.0796'	176.7146'

Practice 49: (Page 127)

	Take out	Center length
(1)	5.3318"	61.3363"
(2)	2.5507"	36.8986"
(3)	19.1832'	111.6347'
(4)	3.7279"	10.5442"
(5)	19.5464"	104.9072"
(6)	0.7367"	15.0266"
(7)	5.3961"	48.2079"
(8)	198'	2244'

Practice 50: (Page 129)

(1) 23" (2) 6" (3) 5.4375" (4) 18' (5) 81" (6) 30" (7) 9.5" (8) 14"

Practice 51: (Page 131)

	Take out	Center Length
(1)	3.2154"	128.1942"
(2)	0.4243'	34.3182'
(3)	0.1786'	3.1428'
(4)	1.6077"	19.5346"
(5)	4.1421'	48.1664'
(6)	2.4853"	54.4264"
(7)	7.4558"	119.4386"
(8)	8.2843'	130.9806'

Practice 52: (Page 133)

	Angle	Take out	Center length
(1)	36.4°	2.6291"	18.3427"
(2)	54.6°	10.3239"	72.5831"
(3)	33.7°	1.8167"	24.3097"
(4)	20.9°	5.5208"	84.4653"
(5)	83.5°	16.0766"	92.2619"
(6)	36.9°	1'	3'
(7)	12.6°	16.6019'	224.022'
(8)	21.2°	4.4902"	43.5603"

Practice 53: (Page 138)

	Angle	Center	Outside
(1)	30°	2.4537"	4.9075"
(2)	10°	0.3527'	0.7053'
(3)	15°	2.1436"	4.2872"
(4)	37.5°	1.1510"	2.3020"
(5)	22.5°	5.3848'	10.7696'
(6)	45°	6.3125"	12.6250"
(7)	28.5°	2.0531"	4.1061"
(8)	62.5°	49.9455"	99.8911"

Practice 54: (Page 142)

	Low Miter	Up Miter	Min. Center
(1)	30.6°	14.4°	101.7094"
(2)	22.5°	22.5°	8.7279'
(3)	22.5°	22.5°	58.8823"
(4)	15°	30°	57.5359'
(5)	18.4°	26.6°	24.1667"
(6)	22.5°	22.5°	96.2584"
(7)	31.7°	13.3°	50.9634'
(8)	36°	9°	60.7611"

Practice 55: (Page 144)

	Miter Angle	Center MB	Hypo
(1)	15°	0.5359"	46"
(2)	22.5°	1.6569"	19.7988"
(3)	45°	2'	8'
(4)	32.5°	3.8224"	37.5148"
(5)	11.25°	0.4973"	177.6925"
(6)	15°	0.6699'	18'
(7)	22.5°	0.9838"	26.1627"
(8)	45°	3'	10'

Practice 56: (Page 148)

(1) 21 in^2 (2) 75 ft^2 (3) 1 ft^2 (4) 43.167 ft^2 (5) 2634.188 ft^2 (6) 10,000 ft^2

Practice 57: (Page 149)

(1) 9 ft^2 (2) 509.07 in^2 (3) 361 in^2 (4) 12100 ft^2
(5) 1369 ft^2 (6) 451.56 ft^2 (7) 4.69 ft^2 (8) 47.84 ft^2

Practice 58: (Page 150)

(1) 54 ft^2 (2) 150 ft^2 (3) 346.5 in^2 (4) 12.92 ft^2
(5) 235 in^2 (6) 3953.61 ft^2 (7) 214.69 in^2 (8) 228.53 in^2

Practice 59: (Page 152)

(1) 62.35 in^2 (2) 18 ft^2 (3) 3.18 ft^2 (4) 0.81 ft^2 (5) 2.84 in^2 (6) 1985.59 yds^2

Practice 60: (Page 152)

(1) 1256.64 ft^2 (2) 31415.93 in^2 (3) 754.77 yds^2 (4) 754.77 in^2
(5) 572.56 ft^2 (6) 42.88 yds^2 (7) 0.79 mi^2 (8) 517.40 ft^2

Practice 61: (Page 153)

(1)	2500 ft^2	(2)	100 in^2	(3)	314.16 ft^2	(4)	63.62 in^2
(5)	1675.21 ft^2	(6)	2450 ft^2	(7)	43.2 ft^2	(8)	89 in^2

Practice 62: (Page 157)

(1)	2376 in^3	(2)	19980 ft^3	(3)	10648 yds^3	(4)	400 mi^3
(5)	9900 in^3	(6)	0.0008 in^3	(7)	7.69 ft^3	(8)	432 in^3

Practice 63: (Page 157)

(1)	1 in^3	(2)	0.000125 ft^3	(3)	10648 mi^3	(4)	1157.625 yds^3
(5)	3.375 ft^3	(6)	970299 in^3	(7)	4492.125 in^3	(8)	1832644.02 ft^3

Practice 64: (Page 158)

(1)	30 ft^3	(2)	630 in^3	(3)	693 ft^3	(4)	4.5 ft^3	(5)	3 ft^3	(6)	2.5 ft^3

Practice 65: (Page 159)

(1)	6.28 in^3	(2)	12.57 in^3	(3)	13.09 ft^3	(4)	903,195.32 yds^3
(5)	105,195.15 in^3	(6)	1,983,179.45 in^3	(7)	94.25 ft^3	(8)	38,641.38 in^3

Practice 66: (Page 162)

	Central Angle	Arc Length		Central Angle	Arc Length
(1)	60°	12.5664"	(4)	45°	124.49 mi
(2)	36°	2.2"	(5)	90°	0.125'
(3)	120°	37.3333'	(6)	30°	6.57'

Practice 67: (Page 163)

	Chord length		Chord length
(1)	9.1844'	(4)	66.7199"
(2)	5.8779"	(5)	1.1705 Mi
(3)	51.7638 yds	(6)	22.5625"

Practice 68: (Page 166)

	Chord length		Chord length
(1)	20'	(5)	15.6198"
(2)	11.388 mi	(6)	3.1058"
(3)	16.9341'	(7)	22.6274'
(4)	64.2908"	(8)	27.6994"

Practice 69: (Page 172)

	Hypo of Δ of roll	Hypo of Offset Δ		Hypo of Δ of roll	Hypo of Offset Δ
(1)	32.2025"	64.405"	(5)	32.1734"	45.5"
(2)	73.0616"	103.325"	(6)	32.1734"	
(3)	67.8012'		(7)	46.7625"	93.5251"
(4)	3.1623'	6.32456'	(8)	5'	7.8103

Index

To order copies of *Math To Build On* or *Pipe Fitter's Math Guide.*

Simply complete and return the order form or a copy of the form below to:

Construction Trades Press
P.O. Box 953
Clinton, NC 28328-0953

MasterCard or Visa orders may be placed toll free by calling:

1-800-462-6487.

Please rush me ______ copies of *Math To Build On* at $22.95 each, plus $2.25 per copy to cover postage and handling. (North Carolina residents please add $1.38 per copy for sales tax.)

Please rush me ______ copies of *Pipe Fitter's Math Guide* at $18.95 each, plus $2.25 per copy to cover postage and handling. (North Carolina residents please add $1.14 per copy for sales tax.)

Name ______________________________

Address ______________________________

City ____________________ State ______ Zip ____________

☐ check or money order enclosed. **Make check or money order payable to Construction Trades Press.**

VISA ☐ Visa ☐ MasterCard **Please fill out information below if using a credit card.**

Total $ amount ________ Card number ☐☐☐☐☐☐☐☐☐☐☐☐☐☐☐☐

MasterCard inter bank #
(4 digit # above name) __________ Expiration date ____________

Signature ______________________________

Prices and availability are subject to change without notice. Please allow four to six weeks for delivery. Remember to add $2.25 per title to cover shipping and handling and North Carolina residents add 6% for sales tax. No cash or C.O.D.'s please. For information on quantity orders call (919)592-1310.